Critical Acclaim for this Book

'This book is essential reading – because of the originality of its point of view, the timeliness of its analysis and the startling propositions which it puts forward.... Bristling with new ideas, the author provides us with an intelligent way into understanding better the world's present disorders.'
Le Monde Diplomatique

'This original and thought-provoking book rejects the widely held belief that development is an inevitable process for nation states. One of its greatest merits is that its author demonstrates that many of the so-called "developing countries" are in fact infected with the virus of non-viability.

He recommends that their politicians, instead of trying to reproduce the present paradigm of prosperity with consumer patterns which cannot be financed and are unsustainable, immediately put in place pacts for national survival which will stabilise their populations and obtain water, food and energy security for their people.

Highly readable, _The Myth of Development_ is a useful tool for understanding the enormous problems that have to be faced by the poverty-stricken, exploding urban populations of the inaptly named "developing countries".'
Javier Pérez de Cuéllar,
United Nations Secretary-General, 1982–1991

About the Author

Oswaldo de Rivero is a former Peruvian diplomat, former ambassador to the United Nations and other international organisations in Geneva (he resigned because of, in his own words, a profound disagreement with President Fujimori's government), and author of *New Economic Order and International Development Law* (Pergamon). His views are the result of a profound knowledge of the international scene acquired during more than twenty years in a broad range of international forums. He represented his country at the UN General Assembly and on the Security Council, in the United Kingdom and the USSR; was president of the Economic Commission of the Non-Aligned Countries' summit; president of the Group of 77 countries; and chairman of the Council of the Latin American Economic System (SELA), a regional intergovernmental body set up to encourage co-operation and integration among Latin American and Caribbean countries. He has also been president of the Review Conference on the Nuclear Non-Proliferation Treaty and the United Nations Disarmament Conference; he led the Peruvian delegation during the Uruguay Round at GATT world trade negotiations. Today he lives in Geneva where he worked as a consultant until his appointment by the new Peruvian government as Ambassador to the World Trade Organisation (WTO) in early 2001. He is currently writing a new book.

A Brave New Series

GLOBAL ISSUES
IN A CHANGING WORLD

This new series of short, accessible think pieces deals with leading global issues of relevance to humanity today. Intended for the enquiring reader and social activists in the North and the South, as well as students, the books explain what is at stake and question conventional ideas and policies. Drawn from many different parts of the world, the series' authors pay particular attention to the needs and interests of ordinary people, whether living in the rich industrial or the developing countries. They all share a common objective: to help stimulate new thinking and social action in the opening years of the new century.

Global Issues in a Changing World is a joint initiative by Zed Books in collaboration with a number of partner publishers and non-governmental organisations around the world. By working together, we intend to maximise the relevance and availability of the books published in the series.

PARTICIPATING NGOS

Both ENDS, Amsterdam
Catholic Institute for International Relations, London
Corner House, Sturminster Newton
Council on International and Public Affairs, New York
Dag Hammarskjöld Foundation, Uppsala
Development GAP, Washington DC

Focus on the Global South, Bangkok
Inter Pares, Ottawa
Third World Network, Penang
Third World Network–Africa, Accra
World Development Movement, London

About this Series

'Communities in the South are facing great difficulties in coping with global trends. I hope this brave new series will throw much-needed light on the issues ahead and help us choose the right options.'
Martin Khor, Director, Third World Network

'There is no more important campaign than our struggle to bring the global economy under democratic control. But the issues are fearsomely complex. This Global Issues Series is a valuable resource for the committed campaigner and the educated citizen.'
Barry Coates, Director, World Development Movement (WDM)

'Zed Books has long provided an inspiring list about the issues that touch and change people's lives. The *Global Issues for a New Century* is another dimension of Zed's fine record, allowing access to a range of subjects and authors that, to my knowledge, very few publishers have tried. I strongly recommend these new, powerful titles and this exciting series.'
John Pilger, author

'We are all part of a generation that actually has the means to eliminate extreme poverty world-wide. Our task is to harness the forces of globalisation for the benefit of working people, their families and their communities – that is our collective duty. The Global Issues series makes a powerful contribution to the global campaign for justice, sustainable and equitable development, and peaceful progress.'
Glenys Kinnock, MEP

A GLOBAL ISSUES TITLE

THE MYTH
OF
DEVELOPMENT

*Non-Viable Economies
of the 21st Century*

Oswaldo de Rivero

Translated by Claudia Encinas and Janet Herrick Encinas

Zed Books
London and New York

University Press Ltd
Dhaka

White Lotus Co. Ltd
Bangkok

Fernwood Publishing Ltd
Halifax, Nova Scotia

David Philip
Cape Town

Books for Change
Bangalore

The Myth of Development
was first published in 2001 by

In Bangladesh: The University Press Ltd, Red Crescent Building,
114 Motijheel C/A, PO Box 2611, Dhaka 1000

In Burma, Cambodia, Laos, Thailand and Vietnam:
White Lotus Co. Ltd, GPO Box 1141, Bangkok 10501, Thailand

In Canada: Fernwood Publishing Ltd, PO Box 9409, Station A,
Halifax, Nova Scotia, Canada B3K 5S3

In India: Books for Change, 28 Castle Street, Ashok Naggar,
Bangalore, 560025, India

In Southern Africa: David Philip Publishers (Pty Ltd),
208 Werdmuller Centre, Claremont 7735, South Africa

In the rest of the world:
Zed Books Ltd., 7 Cynthia Street, London N1 9JF, UK and
Room 400, 175 Fifth Avenue, New York, NY 10010, USA

Distributed in the USA exclusively by Palgrave,
a division of St Martin's Press, LLC,
175 Fifth Avenue, New York, NY 10010, USA

2nd impression, 2003

Cover designed by Andrew Corbett
Set in 10/13 pt Monotype Bembo by Long House, Cumbria, UK
Printed and bound in the United Kingdom by Cox & Wyman Ltd, Reading

A catalogue record for this book is available from the British Library
US CIP data is available from the Library of Congress
Canadian CIP data is available from the National Library of Canada

ISBN 974 7534 762 PB (South-East Asia)
ISBN 1 55266 057 5 Pb (Canada)
ISBN 0 86486 494 9 Pb (Southern Africa)
ISBN 1 85649 948 0 Hb (Zed Books)
ISBN 1 85649 949 9 Pb (Zed Books)

to Juliette

CONTENTS

*

INTRODUCTION

In 1967, at the outset of my diplomatic career, I had the invaluable opportunity of participating in the Kennedy Round of negotiations under the General Agreement on Tariffs and Trade (GATT), where Peru played a leading role among the developing countries, as a major producer of copper, lead, zinc, fishmeal, cotton and sugar. In those years, all of those raw materials were still very important for the industrialised countries. Thanks to that, Peru was able to gain tariff concessions without yielding a great deal in return. Twenty years later, as head of the Peruvian delegation, I again took part in trade negotiations under the GATT. This time the matter at hand was the Uruguay Round, the largest series of trade talks of the twentieth century. Peru's main export products were virtually the same as in the previous round, but this time the negotiations centred on manufactures with high technological content and, above all, on trade in services and on standards for the protection of intellectual property. As a result, the developing countries like Peru, that had neither increased the technological content of their exports in the previous twenty years, nor developed competitive international services, nor invented anything of importance, were virtually left sitting on the sidelines in this, the grandest worldwide trade negotiations of modern times.

After more than two decades of diplomatic experience as a participant in international forums and negotiations, I was the dismayed witness to the gradual loss of Peru's negotiating power. I

was ending my career as the representative of a country that was
archaically inserted in the new global economy, that was still
trapped in the exporting of raw materials or slightly transformed
products with non-competitive prices, that was increasingly
indebted, and that had doubled its population. To make matters
even worse, its strategic situation was becoming critical as it
switched from exporting to importing fuel and increased its food
imports. At the end of the twentieth century, the World Bank classified Peru among the twelve poorest countries in the world, with
more than 40 per cent of its population living on an income of one
dollar per day.

This inability to function in the modern global economy is
hardly an exclusive trait of Peru. The history of the majority of the
countries of Latin America, Africa and Asia, since their independence, has merely recorded a gradual process of dysfunction and
global marginalisation. In this way, a large number of misnamed
'developing countries', undergoing a veritable urban demographic
explosion, are finding it difficult, if not impossible, to modernise in
order to participate in a global economy that demands increasingly
sophisticated manufactured goods and services and that uses less of
their raw materials and their abundant unskilled labour force.

Since the Industrial Revolution and the emergence of the
modern nation-state in Europe and the United States, more than
185 nation-states have been born, most of them in Latin America,
Asia, the Middle East and Africa. A type of historic 'law of diminishing returns of the possibilities of national viability' has accompanied this proliferation across the years. In reality, the majority of
the nation-states that arose in the nineteenth century, like those in
Latin America, and nearly all the new nation-states formed in the
twentieth century, such as the Asian and African states, over a
century later could better be considered as unfinished national
projects that do not develop: quasi nation-states.

Despite having been among the founders in the nineteenth

century of the modern community of republican nation-states, born under the influence of the American and French revolutions, by the end of the twentieth century the Latin American countries had not been able to join the exclusive club of the developed capitalist powers, which currently has just twenty-four members. It has been said that the Latin American countries lost a decade due to the debt crisis, but the truth of the matter is that they have lost fifteen decades, 150 years, without ever managing to become modern, prosperous capitalist democracies. Today, our countries have been overtaken in standards of living and technological modernisation not only by Europe and the United States, but also by Japan, Taiwan, South Korea, Malaysia, Hong Kong, Singapore, Australia, New Zealand and Canada. In the nineteenth century, when Latin America made its historical debut, independent and rich in natural resources, those nations were either very poor, semi-feudal countries, or not very prosperous British colonies. The socio-economic landscape of Latin America 150 years ago resembled a European province or the North American frontier. By contrast, today it looks more like the poor countries of the Middle East or Asia. In less than a hundred years, Europe and the United States succeeded in eliminating virtually all their poverty, while in Latin America poverty has become practically hereditary.

Nevertheless, the technological lag and the poverty of Latin America seem insignificant compared with many African, Asian and Middle Eastern countries, where tribal, ethnic and religious differences fracture cohesion, complicated more recently by the spread of Islamic fundamentalism and its rejection of modernity. Many African, Asian and Middle Eastern states that emerged in the middle of the twentieth century have not been bogged down in underdevelopment, like the Latin American nations. They did not even experience the booms enjoyed by the latter with their exports of guano, saltpetre, rubber, coffee, sugar, cotton, wool or minerals. By contrast, these African and other states emerged without any national development options, due to the unfortunate coincidence of their independence with a technological revolution that needs less and less of the raw materials and the abundant supply of manpower that are their only comparative advantages.

In the majority of the industrialised states, national identity preceded the formation of the state authority. The nation, reflected above all in the joint emergence of a middle class and a market of national dimensions, formed the base of the modern state. In contrast, in most of the so-called developing countries, this sequence was reversed. The political authority – the state – emerged from the independence process before the nation, that is, before the development of a true bourgeoisie and a unifying national capitalist economy. For this reason, the majority of the wrongly termed 'developing countries' are children of their enthusiasm for freedom, but not the offspring of middle-class prosperity and scientific and technological progress. It has not been possible to replicate the developed, capitalist and democratic nation-state in most of the countries that comprise the so-called developing world. The greater part of humankind continues to exist with low incomes, in poverty, technologically backward and governed by authoritarian regimes or, at best, in low-powered democracies.

At the end of the twentieth century, the world really consisted, aside from the 24 developed counrties, of more than 140 non-developed countries and of only 4 developed 'newly industrialised countries' (NICs): two city-states (Singapore and Hong Kong) and two small countries (South Korea and Taiwan). These constitute only 2 per cent of the population of what the experts have been calling, for the last forty years, the 'developing world'. The four are the only cases in which, despite the financial crisis of 1997, it is possible to verify a significant technological modernisation of production and of exports, a continuous process of income redistribution and a significant shift of population from the poor to the middle class, nearly comparable to what happened first in the United States and in Europe, and later in Japan. In spite of such progress, these NICs are far from enjoying the scientific, technological and cultural development and the standard of living – and, even less, the democratic development – of the United States and Europe.

During the Cold War, many of the unfinished national projects, euphemistically called 'developing countries', acquired strategic value by taking advantage in one way or another of the East–West conflict. This provided them with room for manoeuvre, enabling them to obtain economic aid and political support from one of the two power blocs, and to finance their economic non-viability in this manner. This strategic subsidy allowed many countries to survive despite profligate economic policies and excessive state intervention, and it allowed them to indulge in extravagant dreams. The end of the Cold War turned those dreams into nightmares. Today, under the supervision of the IMF, the World Bank and the World Trade Organisation (WTO), all these countries are obliged to integrate into the global economy, where to their misfortune the greater part of them will be unable to withstand the competition. They will be sidelined by the Darwinian operation of the global economy and its technology.

The globalisation gurus, however, are convinced that the prosperity and the development of all countries will be achieved this time, as a result of worldwide competition within a totally unfettered global market. Such a belief, with its utopian ingredients, presents globalisation as an unstoppable process, beyond human control, as though it were the universal law of gravity – inescapable for persons, for enterprises, and even for nations. What the globalisation gurus do not mention is that the United States, Europe and Japan would certainly not have managed to develop under this sort of globalisation. During their development processes, they protected and promoted their young industries and copied each other's technology. What is not explained, either, is that today's global economic arena is not as unfettered as purported, because it neither allows the free movement of human beings in search of work, nor permits copying of foreign technology, as was formerly the case. The workers and the technologies that are so vital for the underdeveloped countries are not permitted to circulate globally. They fall under the strict regulation of immigration and intellectual property protection laws.

All of this does not mean that globalisation is a completely negative factor. Thanks to this process, national cultures are brought closer together, and political, economic, and scientific information, as well as amusements and sports events, are disseminated throughout the planet. In addition, with the benefits of global communications, today no regime can hide the human rights violations committed against its citizens. No one would wish to deny the benefits derived from globalisation. Rather what is criticised is its speculative nature, which has converted the international financial system into a kind of global casino, where mega financial crises are spawned, such as those that struck part of Asia as well as Russia and Brazil in the late 1990s. In addition to the tendency towards speculation, which contributes nothing to production, or to employment, another worry is the growing use of labour-saving technologies, precisely at a time when the underdeveloped world is producing an urban population explosion.

Today the technology applied in the globalisation process is creating social exclusion, because computer software and automation are saving on human labour. As the twentieth century came to an end, 30 per cent of the world's working-age population were unemployed. The enormous factories with their chimneys and the large proletarian populations are being gradually phased out by new technologies. In many industrialised countries, this situation can be tolerated, because the population is not growing and the services sector can often absorb the manpower that has become redundant in the manufacturing field. However, in underdeveloped countries, where the urban population explosion increases the supply of unskilled labour, it will be next to impossible for the new technologies to create a sufficient number of jobs. Thus, the technological revolution is starting to enter on a collision course with the demographic explosion in the poor societies. The new perversity of the global economy is starting to take shape as gross national product (GNP) grows hand in hand with unemployment or the numbers of badly paid temporary jobs. The new globalising technologies are also beginning to segregate the raw-material-exporting economies from the industrialised economies. The quantity of raw material required for each unit of industrial production continues to diminish, thereby destroying the chances of development for those exporting countries. Now, at the beginning of the twenty-first century, the amount of raw material per industrial unit is only 40 per cent of what it was in 1930. While the global economy today requires increasing amounts of high technology manufactures and, above all, sophisticated services, it needs less and less primary, or only marginally transformed, products.

The only comparative advantages of the underdeveloped world – abundant manpower and raw materials – are becoming every day less important to the global economy. Within a globalising process of this sort, how can primary production countries, where the demographic explosion is adding millions of youths to the labour market each year, become economically viable? Undoubtedly, with this trend, it will be very difficult significantly to increase income per capita, provide jobs for 700 million unemployed workers, and extricate from their extreme poverty 1.6 billion inhabitants of the underdeveloped world. Development no longer depends on democratic national efforts and decisions. Global trends and powers not elected by the people will be the determining factors in the national destiny of many countries.

In addition to the impasse that globalisation's present economic orientation creates for the backward, primary-goods-producing economies with abundant labour, the patterns of consumption that it is promoting are by and large unfriendly to the environment. They are creating chaotic planetary urban expansion, causing declining yields in the most fertile agricultural lands, diminishing the supply of fish, increasing the processes of deforestation, water scarcity, and desertification, and affecting the climate with their emissions of greenhouse gases into the atmosphere.

As we begin a new century, we enter *terra incognita* for those countries that did not manage to increase the technological content of their production or to stabilise their populations during the twentieth century. How can the underdeveloped countries attract foreign investment and technology, in order to escape from the trap of primary, or barely transformed exports, when transnational capital has no interest in investing in new, modern industries in the majority of these countries? How can jobs be provided for hundreds of millions of workers in the underdeveloped world, while employing modern technology that tends to save on human labour? Within these trends, what can be done to convert nearly 5 billion inhabitants of the underdeveloped countries into a planetary middle class, with sufficient income to be integrated as consumers of global capitalism? The challenge raised by these unknown factors is immense, but even were it to be met, another one immediately arises: how could the 5 billion inhabitants of the underdeveloped world adopt the consumption patterns at present shared by only 1 billion inhabitants of the advanced capitalist societies, without causing a real environmental catastrophe?

Facing these challenges of the twenty-first century, Latin America, Asia, and above all Africa are rife with unsuccessful national projects. Their massive poverty prevents them from building nation-sized market economies, and the lack of technological input in their manufacturing and service enterprises prevents them from competing in the global economy. Soon many of these so-called 'developing' countries, plagued by the urban population explosion, will begin to show signs of economic non-viability, due in part to the absence of a national and an international market. Their only function as part of the present global economy will consist in paying their debt, receiving some speculative foreign capital, and importing food, fuel, and all sorts of industrial and consumer goods.

For a great many poor countries, the option during the coming decades will not be to embark on a development process, as occurred with South Korea or Taiwan twenty years ago. Their only hope will be merely to survive, in some manner, the challenges of the technological revolution and global competition. This fact may seem shocking, since until now it has always been assumed that every country has the capacity to become developed. The experience of the twentieth century, however, indicates the contrary and obliges us to think the unthinkable: that many of the wrongly named 'developing countries' are not on the road to becoming newly industrialised countries (NICs); instead, they are slipping towards the status of non-viable national economies (NNEs). If their situation should worsen, they could implode into violence, as ungovernable chaotic entities (UCEs), as has already happened with some countries of Africa, the Balkans, Asia and Latin America.

After nearly half a century of theories, forecasts, promises and searches for El Dorado, this book deals with the danger faced by a sizeable group of African, Asian and Latin American nation-states: of becoming bogged down in non-viability. I am aware that this subject represents a kind of taboo in very poor, socially disintegrated countries, where nationhood has not been consolidated and where democratic traditions are scarce. By voicing truths that have been studiously avoided, we run the risk of causing deep discomfort and of injuring false patriotic sentiments, which have only served to mask the historic dysfunction of the nation-state. I am convinced, however, that such reflection is essential in countries that have wasted the entire twentieth century without achieving development, where this process is still elusive in the present global jungle, and where the illusion of development has to be replaced by a realistic pact for survival.

CHAPTER 1

THE TWILIGHT OF THE NATION-STATE

Quasi nation-states

Seen from outer space, our planet appears as a blue orb, robed in a thin film of life, the biosphere. Inside that layer, microorganisms, plants, animals and the human species exist. By dint of centuries of violence and political evolution, the latter gradually organised the earth's territory into different nation-states. Although these entities' frontiers are invisible from outer space, they are ever present here on earth. With the exception of the polar regions and the oceans, not one centimetre of the planet exists without delineation and occupation by some state authority. At the end of the twentieth century, there were over 185 nation-states, and that number may still increase, with time. This form of political organisation continues to constitute the ideal for numerous human communities aiming to differentiate themselves from other groups, to achieve security and prosperity, and to participate on the international stage as sovereign nations. Throughout its history, humankind has given shining examples of heroism, of altruism and of creativity in the name of the nation-state but, in that same name, it has perpetrated acts of cynicism, cruelty, human destruction and environmental waste.

The nation-state, as we know it today, is the product of four hundred years in the evolution of Western political thought. Its foundations hark back to the Renaissance theses about the reasons for the existence of the city-states put forward by Niccolo Machiavelli, and, above all, to the ideas of Thomas Hobbes. Hobbes expounded the most convincing arguments of his time concerning the necessity for a supreme central authority in order to liberate man from his natural, brutish state, and grant him security. Hobbes compared this highest authority to the Leviathan, the supreme biblical monster described in the book of Job, whose power was unparalleled. From that time forward, the Leviathan became the idol of a new civil cult exalting the 'reason of the State'. In its name, mountains of human sacrifices have been offered. The cult of the Leviathan has encompassed a great variety of rituals, from absolute monarchy to democracy, passing through nazi–fascist and communist totalitarianisms on the way.[1]

The absolutism of European monarchs was the human incarnation of the Leviathan. During the sixteenth century, the monarchs extended their reign over feudal lords, counties, duchies, free cities, and in general over all the feudal powers of that time. They imposed a recruitment method for the royal armies, applied a centralised system of tributes, minted money, created the public treasury, and established the nucleus of what would become modern state bureaucracy.[2]

The continual fighting under royal flags and emblems, the hegemony of a common language over Latin and the existing dialects, as well as the adoption in all the kingdoms of Europe of the Christian religion, in its Catholic or Protestant versions, all combined to increase each population's identification with the monarchy and to fortify the state, lending it the significance of the present-day nation-state. In 1648, the Treaty of Westphalia, which put an end to the wars of religion under the European monarchs, established the classic characteristics of the modern nation-state, closely patterned on the attributes of monarchy. Since that time, states have been seen as sovereign and equal, as were the kings before them. There is no authority or entity above them. All are Leviathans and, as such, are supreme, sovereign, equal and independent powers. Somewhat later, Louis XIV of France and Frederick the Great of Prussia personified this absolute sovereignty, with enormous bureaucracies and great military power.

With the independence of the United States in 1776, the monopoly of sovereignty held by the monarchies began to disintegrate. That revolution laid the foundations for the cult of the state under republican, democratic procedures and the respect for the individual's civil and political rights. In 1789, the French Revolution adopted the American principle of guaranteeing individual freedoms. However, instead of investing sovereignty in the people, as decreed by the United States Constitution, it placed sovereignty in the hands of the 'nation', a new, abstract concept born of French rationalism. The Declaration of the Rights of Man and of the Citizen, of 1789, proclaimed that no individual could exercise any sort of authority that did not emanate from the nation. But what was the nation? According to Sieyes, the nation was nothing but the third estate, or the general will of the majority, as Rousseau had propounded.

The French revolutionaries could not have imagined the totalitarian consequences that might derive from the interpretation of this idea of the general will. In fact, the Jacobin revolutionary terror shortly thereafter proved very proficient in interpreting the general will and representing the nation above individuals, especially if these individuals were aristocrats or enemies of the Jacobins. Thus it was that, paradoxically, the exaltation of the nation allowed the Leviathan to increase its power and to override the human being's individual rights. Consequently, it is not surprising that, from that time on and throughout the ensuing pages of history, totalitarian interpretations should arise, confusing the general will of the majority or of the 'nation' with that of a predominant ethnic group or a predestined social class. The Nazi state and the Soviet state were perverse results of the personification of the general will in the Aryan race or the proletarian class. Ideologies such as Nazism or communism, perhaps inspired by Rousseau, were very distant from Jefferson, whose main concern, following Anglo-Saxon tradition, was the protection of the individual's inalienable liberties against the Leviathan's excesses or excesses of any other political abstraction, such as the 'nation'.

Without a doubt, it was the Industrial Revolution in Europe and the United States that put the final touches on the modern nation-state as we know it today. The development of industrial capitalism identified the cult of the Leviathan with the creation of a national market and a beneficial integration into the international market. The paradigm of a nation-state that was sovereign, integrated and united – not only by ethnic, cultural and religious ties, but also by the material well-being of its population – prospered in various parts of the globe. To the Leviathan cult was added the concept of national economic progress. In this way, the new civil religion, originated with Hobbes, was brought to its completion with the prediction that personal prosperity and happiness would be achieved through the growth of the nation-state's GNP. Thus were born the twin myths of progress and development, which still today are pursued as El Dorado by the majority of the backward and underdeveloped countries which have never undergone a real capitalist industrial revolution.

The illusion of a republican and democratic nation-state, where the people's well-being and happiness would be assured, was fundamentally the product of the American and French revolutions. After that era, it began to take root all around the world. In the nineteenth century this idea finished off the Spanish and Portuguese empires, giving rise to the new Latin American republics. At the beginning of the twentieth century, as a result of the First World War, the ideal of the nation-state destroyed the multinational Austro-Hungarian and Ottoman empires, and gave rise to new states in the Balkans and the Middle East.

After the end of the First World War, the dream of having a state of one's own grew ever stronger; this was as a consequence of the principles proclaimed by Woodrow Wilson, and confirmed in the Versailles treaties, concerning the right of nationalities to create their own state organisation. Wilson's misguided idealism awoke the dragon of nationalism in all its guises. Starting with Versailles, every human group endowed with some ethnic, cultural and religious affinity felt that it had the right to become a state, even though it did not constitute a true nation and did not have the economic and technological means to be viable. Thus the cult of the Leviathan had reached its apex.

The nationalist dreams of the twentieth century relied on the principle of self-determination as their political and juridical instrument. Its application so far has been based on the assumption that as many nation-states can be created as there are nationalist elites that wish it, with no thought for these new states' governability or viability. The only thing needed is international recognition. While independence admittedly gave dignity to peoples who had been the victims of domination and discrimination, it did not necessarily create viable nation-states. The result of this is that a large number of countries find themselves in a worse situation than when they were colonies, and many of them wish they could be recolonised.

The cult of a Leviathan of one's own and an ideology based on the principle of self-determination caused an unprecedented proliferation of nation-states during the Cold War. At that time, demagogues scoffed at any caution in applying the principle of self-determination, treating it as a pro-colonial, imperialist or racist attitude. To delay the right of self-determination unleashed the counterpart right to wars of liberation with the accompanying duty to help the insurgent population. It was anathema to go against the decolonising avalanche that tried to reproduce the European model of the nation-state in human communities that had no concept of the state, or of the nation, and that lacked both the middle class and the national market they needed in order to be governable and viable. Upon granting them recognition as independent countries, the rivals of the Cold War lost no time in lavishing international aid on them so as to exercise their own influence on the new nation-states. When the Cold War ended, the strategic value of these countries evaporated, leaving them on their own, virtually without aid or special treatment as developing countries. They were at the mercy of a process of natural selection on the part of the new global economy, which needs less and less of their raw materials and abundant labour force.

The principle of self-determination of the United Nations Charter was applied during the decolonising fever without concern for the political, economic, social and cultural factors that determine the governability and the viability of a nation-state. Decolonisation within the United Nations became a rather routine diplomatic posture to avoid making waves during the Cold War. This stance prevented a calm and gradual application of the self-determination principle, an application that would take into consideration the possibility of instituting a process toward self-government and economic viability. The colonial powers seemed to be in a great hurry to rid themselves of the explosive socio-political burdens caused by an anti-colonial movement that was more fired up by nationalist ideology than by the feasibility of economic and social development. Even more, the ideological embodiment of self-determination reached such heights of fantasy as to believe that it was impossible to have development without independence and that, in the end, it did not matter that a country be born poor, since international aid would bridge the economic gap with the former metropolis. Today's reality stands in stark contrast. Economic and social development is merely a distant myth propagated by the political classes and international technocracies in these poor countries. After fifty years of experiments in development and billions of dollars in aid, the majority of them are not developing, but rather are becoming more and more under-developed.

The emancipation euphoria often propelled by tribal nationalism and the Kalashnikov has ended in catastrophic processes of under-development and national non-viability. The uncontested dream of one's own Leviathan overrode the real possibility of many human communities to organise themselves as civilised states. The majority of the member states of the United Nations supported this illusion, often with ideological automatism, without measuring the later consequences on regional and world stability to be caused by independence devoid of economic viability. In applying the principle of self-determination, they did not take into account the minimum prerequisites for the governability of the new entity, its capacity to provide well-being for its population, the availability of competitive enterprises, technology, food and energy production, as well as the probability of its exercising respect for human rights. Dozens of states joined the hitherto exclusive Leviathans Club, without having the conditions for their own future governability and viability. They were recognised as sovereign, but paradoxically were considered in need of inter-national aid in order to survive. In direct contrast to the nature of the Leviathan, they were recognised as 'unequal' states. In other words, they were seen as 'incomplete' *quasi nation-states*, 'needing to develop'. Time would prove that they would never be com-pleted either as states or as nations, and that the majority of these underdeveloped entities are not Leviathans. The idea that the European model of the nation-state could be reproduced proved to be not only false but dangerous for the stability of the region and of the world.

During the Cold War, all those false, incomplete Leviathans called developing countries acquired strategic value by taking advantage of the East–West conflict in one way or another. Thus they gained room for manoeuvre for the purposes of obtaining economic aid or political support from one of the two blocs, in order to finance their non-viability. This allowed the dream of the nation-state to continue and entities that lacked future viability to survive. The end of the Cold War has turned that dream into a nightmare. Today the governments of the so-called developing states are beginning to wake up and to confront the cruel reality of their population explosion, lack of a national market, meagre production of food and fuel, and their low-priced raw materials exports. In addition, they lack a strategic position which would permit them to negotiate more aid, a reduction of the heavy payments on their foreign debt or a 'special and differentiated treatment' in trade, investments or intellectual property. At present, with the supervision of the International Monetary Fund (IMF), the World Bank and the World Trade Organisation (WTO), all these nations are obliged to take part in the global economy on equal terms with the industrialised countries. A great majority of these poor, technologically backward countries will be unable to stand the transnational competition and their national capitalist enterprises will be absorbed or discarded as inappropriate economic species.

In the end, the price for the thoughtless overuse of self-determination in the second half of the twentieth century, together with the loss of the underdeveloped countries' strategic importance, is being paid by millions of unemployed young people in the countries that became independent over that period. Now they think only of emigrating to the capital of the former colony against which, ironically, their fathers and grandfathers had rebelled so as to give them a nation-state. It is not so strange, therefore, that the inhabitants of Puerto Rico and of the Pacific island of Palau do not want to become independent from the United States and that the inhabitants of the Comoros wish to be recolonised by France.

Nowadays, support for the right to self-determination is not as enthusiastic as it once was and is tempered by worries about the fragmentation processes occurring in multinational nation-states such as the Soviet Union and Yugoslavia. The great Western powers, which had the responsibility to create a new international order after winning the Cold War, have done very little to preserve Yugoslavia's unity or the new version of the territorial economic union of the USSR proposed by Gorbachev. This inertia in the face of the disintegration of such strategic states may carry a very high price in the future. This would certainly be the case if the capitalist democratic project in Russia were to fail, or if new fighting broke out in the Balkans, or if the new quasi nation-states that arose from the former Soviet Republics in Central Asia and the Caucasus were to fall apart. As the USSR and Yugoslavia fell into fragments, new Caucasian, Central Asian and Balkan states were recognised, even though they had no experience in self-government, were steeped in tribal nationalism, and had precious little capacity for survival as states in the twenty-first century.

In the majority of the industrialised states, national identity preceded the crystallisation of the state authority. In other words, the nation, reflected in a common culture, and above all in the emergence of a middle class and a national market, existed before the modern state was formed. In contrast, the majority of the quasi nation-states of Latin America, Asia and Africa, despite their historical and cultural differences, experienced this sequence in reverse. The political authority, that is to say, the state, emerged before the nation, before the national cultural identity and before the development of a true middle class and a unifying national market. As a consequence of this, in many of these countries the political elite, the state bureaucracy and the military are still trying to achieve a national project, through the use of symbols and pursuing myths that serve them as sustenance.[3]

Throughout the twentieth century, the elites of the underdeveloped countries have wanted to reproduce the modern European or North American nation-state or, in some cases, have tried to copy the Soviet model. Nearly all of these attempts have ended in disaster. Not to allow imitation would seem to be an irony of imperialism. The underdeveloped countries' elites, through a variety of national projects, have pursued the myth of development. This myth took on the shape of state intervention or of a socialist revolution, and is now in the guise of a neoliberal capitalist revolution. In all these cases, the authorities have exacted mountains of social sacrifice, without managing to eliminate poverty and establish a true civil society ruled by law and by democratic institutions. The cost of the Soviet version of development was shortages and lack of freedom; today, that of the neoliberal, capitalist variant is unemployment and social exclusion. For the great majority of the so-called developing countries, it is increasingly difficult to achieve the formation of a nation-state united by a national market, high standard of living and individual freedom. The global socio-political conditions in today's world make it very difficult to repeat the experiences of such former British colonies as the United States, Canada, Australia and New Zealand, which are the only former colonies where the large majority of the population enjoys both a high living standard and freedom.

One of the clearest characteristics of the quasi nation-states of Latin America, Asia and Africa is the lack of connection between the official world and the vast ocean that constitutes the semi-urbanised population. This human mass organises in its own manner, ignores legal and other formalities, conducts a separate economy that does not appear in the national accounts, and overwhelms the state with its demands and its spontaneous organisation. This population is largely unemployed or underemployed, living outside the national and the global consumer society; it has recent rural roots and is partially urbanised, with no real awareness of nationhood. It often attempts to affirm its identity, not as a social class, which it is not, but rather through ethnic or provincial affinities, ancestral myths or in religious–magic interpretations and radical ideologies. These may even grow into cultures that violently reject modernity, as is the case with the various strands of

Islamic fundamentalism, the Khmer Rouge, the Tamil Tigers or the Shining Path, as well as other radical movements that are emerging in countries where the state does not have an integrated nation at its base.

The appearance of quasi nation-states poses a novel fact for the theory of international relations. Since the emergence of the modern state, despite legal pronouncements about equality, there have always been powerful nations and weak nations, large and small. In the nineteenth century, however, the smaller and weaker states, such as Belgium, Switzerland, Holland, Denmark or Japan, managed to develop through their own efforts, with some help from occasional allies. In the twentieth century, the quasi nation-states have been stabilised in underdevelopment for many years, and are surviving in part because of international aid. This means that they are not viable with their own resources.

How can the quasi nation-states be made economically viable when their populations are growing explosively and their exports consist of primary or only slightly processed products, which fetch low prices and are in little demand? How are we to deal with ungovernable countries where corruption is rife and the daily practice of democracy is rudimentary at best? How are the market economy and consumer society to be produced in Latin American, Asian and African countries that have more than 40 per cent of their population living below the poverty line, on less than one dollar a day? How are nearly 5 billion persons with low incomes to be integrated into global consumption patterns, without seriously damaging the biosphere? How is the enormous gap between rich and poor countries to be closed without gravely affecting the planet's ecological balance?

The myth about closing the gap between the so-called developing countries and the industrialised nations has translated into a splendid disaster. Three decades of United Nations efforts in favour of development have resulted in a kind of world socio-economic apartheid: a planet in whose northern hemisphere there is a small archipelago of wealthy nation-states, surrounded by the majority of mankind. The latter comprises the population of more than 130 poor, or extremely poor, quasi nation-states, where the government does not control economic life, where the state is totally absent from entire provinces, where the urban population is exploding and the majority lives in the informal sector, where life is tumultuous and difficult, and where emigration is the only way out for the young.

These quasi nation-states that cannot develop lack the essential attributes of a modern nation-state. They do not have market economies of national dimensions, because of the numbers of inhabitants that live in poverty or below the poverty line. Besides, they do not control large segments of their economic activities, because these are mainly informal, and what remains in the formal sector is controlled by the IMF and the World Bank. Nor do they have jurisdictional control over all their territory, since large areas are in the hands of insurgent groups, bandits or drug dealers. And in many quasi nation-states, political life itself is controlled from abroad, with external monitoring of their human rights obligations and of their questionable electoral processes.

In the international field, the quasi nation-states have no nego-tiating power and do not exercise a positive influence on any major event. Instead, they are often the source of problems for the international community. They appear in the world press as terri-tories with elected, but not democratic, governments, lacking basic institutions, where barbaric acts occur and human rights are violated, where armed confrontations and drug-driven terrorism take place, or where governments are violently overthrown. Another characteristic of these entities is their inability to be partners or allies, as a result of their weakness. The central activity of their foreign policies, if such a thing exists, is to solicit aid and exoneration from their international obligations, to accept economic adjustment programmes and periodically to restructure their foreign debt.

Most of these quasi nation-states exercise a kind of negative sovereignty, since they do not have the supreme power to achieve well-being and security for the majority of their population. Nevertheless, in some cases, they make a public display of their sovereignty, invoking the right to 'non-intervention in internal affairs' when the international community demands that they comply with their international obligations in matters of human rights. Still, these quasi nation-states have even been expelled from this last trench of their negative sovereignty by the international monitoring of human rights violations and by selective actions of 'humanitarian intervention'.[4]

Perforated sovereignties

All nation-states have been principally engendered by revolution and war. For that reason, their security is based more on the military than on any other factor. War made the nation-state the supreme actor in international relations. The Leviathan was the only entity capable of changing an international situation from pacific to warlike, with all the consequences a decision of that nature had on the lives of its citizens and of its enemies. The nation-state's high points in the role of master of lives were the two world wars of the twentieth century. Starting from the Cold War, the nation-state's role as supreme actor began to decline because, for the first time, nuclear parity impeded the most powerful Leviathans from making war in order to defend their national and ideological interests.[5]

The balance-of-terror policy, called MAD (mutually assured destruction) prevented an Armageddon, originating a period of strategic stability that lasted more than forty-five years. This facilitated the global expansion of capitalism and the entrance on the international stage of new, non-state actors, the transnational corporations. During the Cold War, the number of these enterprises burgeoned, from 7,000 at the beginning of 1960 to 37,000 at the end of 1989, thereby producing a second scientific and technological revolution as important as the Industrial Revolution, if not perhaps even more significant. Not only did the transnational enterprises integrate the economies of the United States, Europe and Japan, but they connected them with the entire world, including even the economies of their Soviet bloc rivals. Ideological confrontation and the arms race during the Cold War did not stop the transnational corporations from doing lucrative business with the Soviet bloc. They managed to circumvent many prohibitions about investing and selling technology to the Communist countries. They opened branches of both Western banks and Western companies in the capitals of those countries; their investments in joint ventures with the countries of the Soviet bloc reached billions of dollars. The most graphic demonstration of the transnationals' relations with the Soviet bloc was the construction of the gas pipeline from Siberia to Europe, producing Warsaw Pact gas for NATO, no less.

As the Cold War faded, the transnational corporations continued to make inroads into the sovereignties of all nation-states. Today the greater part of the goods, services, financial transactions, entertainment and publications is produced by transnational enterprises. In a modern world that is becoming global through the action of these powerful enterprises, states have been losing sovereign control over economic and cultural decision-making. Globalisation is eroding national capitalism, which constituted one of the foundations of the modern nation-state.

Today, the threat to the Leviathan's sovereign power is not an invasion by foreign armies; it is rather the world scope of the economy that allows decisions taken outside the national territory to determine the behaviour of interest rates, the fiscal deficit, the currency value, the price of primary products, the amount of unemployment, or the relocation of entire industries. Activities that were formerly reserved as strategic have practically disappeared. They may be taken over by companies that are located abroad, and even in states that were traditionally considered rivals. At present, even the arms industry of the one remaining superpower, the United States, is globalised. Various of that country's arms systems depend on the manufacture of parts and on technologies produced by companies that are not based in its territory.[6]

Even the richest and most powerful states nowadays often try to coordinate their national policies with other states in order to solve such problems as unemployment and maintaining the value of their currencies. That is what happens at meetings of the Group of Seven most industrialised countries. For more than twenty years, the heads of state of the United States, Germany, Japan, France, Britain, Italy and Canada have met regularly in an effort to manipulate the world economy and to find solutions for global problems, but these meetings have been without visible results. Their attempts at economic coordination have been upset by global financiers: not even these powerful countries have a sufficient volume of reserves to defy the global speculators.

The globalisation of the financial world is one of the transnational phenomena that has most potently impaired the nation-state's sovereignty, causing it to lose control over its own currency and fiscal policies. Today, the international financial system resembles a huge casino. Speculation amounting to billions of dollars occurs daily in the foreign exchange of the world's five wealthiest states, without any possibility of them being able to exercise control. A change in the value of one currency with relation to another can cause bankruptcies or bonanzas, inflate costs, produce unemployment, stimulate imports. All these bets are placed electronically through computing and telecommunications, at the speed of light, by international brokers in New York,

London, Tokyo, Frankfurt, Paris and Singapore, all of them beyond the purview of even the Leviathans' finance ministers and central banks.[7]

The development of telecommunications and information by means of global television has brought into contact the most diverse nationalities and cultures. It has spread all over the planet the image of a Western style of living, based on high consumption, material comforts, and permanent entertainment (music, films, videos). However, this cultural penetration is not causing a parallel global distribution of the democratic values and the respect for human rights that are the very substance of Western civilisation. In the new generations, a sort of cultural homogeneity is being created. Its attraction consists in its promotion of the instant gratification of material needs, from sex to fashion. However, it is not necessarily creating a new planetary ethic, since it neither fosters human solidarity nor promotes environmentally friendly patterns of consumption. In today's world, in sum, capitalist comfort can live side by side with barbarity.

Nowadays, no state can isolate itself from the seductive transnational images that give priority to instant individual gratification over equality and solidarity. In its most radical individualistic version, this capitalism appears as the only paradigm for the search for happiness in every country. It is accepted and desired, despite the risks of social exclusion, because hopes are high that some day one will partake in the material life's banquet. In reality, however, the consumer society does not extend globally for the estimated 1 billion unemployed that exist at present around the world. In these conditions, it is not surprising that violence and fundamentalist movements are arising as a result of the frustration felt at not having reached the levels of consumption that are globally advertised as a possibility for all.

Despite these trends, however, globalisation of the communication media does perform a positive role in unifying humankind.

Not only does it transmit images of the Western materialist lifestyle, but it also transmits images denouncing massive violations of human rights, abuses and injustices in the world. This is building up a common feeling around the world of real concern for human problems and sufferings. Men and women are beginning to realise that they are part of humanity and not only citizens of a single country. This new global human conscience cannot be controlled by any one state, no matter how powerful it may be. Today the sovereignty of nearly all nation-states has been penetrated by the global revolution in telecommunications. The way in which governments treat their citizens is observed by the whole world.

The process of transnationalisation and globalisation of the economy has gone hand in hand with an unprecedented scientific and technological revolution that is creating incredible opportunities for prosperity, although it also raises colossal obstacles for the underdeveloped quasi nation-states. Industrial production today requires increasingly fewer raw materials and energy per unit of production, due to the invention of substitutes, new artificial materials and computerised organisation.[8] Many transnational corporations are capable of producing, in the laboratory, agricultural products that the underdeveloped countries traditionally export. In the same way, they are creating artificial materials that replace metals. This new technological trend will doubtless affect the viability of the economies of so-called developing countries to such an extent as to leave them virtually producers of the obsolete.[9]

The transnational technological revolution cannot absorb the 47 million persons who annually enter the labour market around the world. The competition among the transnational corporations forces them to automate their plants and restructure their production methods, frequently creating more unemployment than employment. The promise of full employment as an objective of the nation-state is today unattainable, and large sectors of their populations are unemployed and excluded from society. In the quasi nation-states, unemployment will take on unforeseeable proportions in the years to come, because advances in technology and in automation will coincide with urban population explosion.[10]

The critical environmental problems of the planet also undermine the sovereignty of the nation-state. When estimating the wealth of the nations, present-day national economic policies do not deduct from GNP the nations' irreversible ecological losses. In this way, resources are exploited until total depletion. At the same time, these policies foster consumption patterns that destroy the environment. These consumption patterns are very difficult to change, since to do so would create the risk of great social turmoil: high-income citizens would not wish to relinquish their high living standards and the poor would not like to give up their dream of some day living like the rich.

This impasse can only be gradually solved, as a supranational environmental management system begins to emerge, with the participation of states, the transnational corporations and representatives of civil society. This system would lay down measures and provide financial and technological resources for sustainable economic activities and management of the common heritage of mankind, in order to avoid irreversible damage to the biosphere.[11] An embryo of this system can be observed in the present regimes to control the ocean depths and to care for the resources of Antarctica. If this embryo of a supranational system were to be extended to other areas of the human heritage, even the powerful industrialised nation-states would become administrators of supranational standards that would be applied in their own territories. The nation-state of the twenty-first century would then be very different from the powerful Leviathan that produced two world wars in the twentieth century.

Powerless powers

Today we are confronted with a sort of 'Law of Diminishing Returns of National Power'. The majority of the states that became independent in the nineteenth century gradually lost what little power they had and those that were freed in the twentieth century were born with practically no national viability. The Latin Americans belong to the first group. In the second group are found the great majority of the Asian and African countries. The partial exceptions to this law are China, India and Pakistan, as they have acquired nuclear power, although the bulk of their population remains trapped in poverty, and two newly industrialised countries (NICs) of Asia, Taiwan and South Korea, which have acquired economic and technological power.

The power in the world has not been redistributed in more than a century. During the last hundred years of history, the most powerful states have nearly always been the same. Britain, Germany, Japan, France, Italy, Russia and the United States were already powerful in the nineteenth century. Not one of the largest countries of Latin America – Brazil, Mexico and Argentina – in spite of having been founders of the nineteenth century community of modern republican states, has been able to gain entrance to the club of the great powers. In fact, most of the countries in the world are losing history's marathon. Not only have they not managed to develop and participate in world power, but many of them have been losing national viability in the face of the enormous challenges that the global economy and the technological revolution present.

But the new twist is that nowadays not only most underdeveloped states are losing national power. The great Western powers that had maintained themselves as a controlling oligarchy during the last hundred years do not have sufficient power, either, to organise the world. At this juncture, there is no community of great powers with the capacity for creating a new world order, as was done in Vienna in 1815 and in Yalta in 1945. Today nobody can put the world in order.

The United States is the only global military superpower today, because it has the capacity to send its military forces to any part of the globe. Paradoxically, however, it does not have sufficient capacity to occupy territories and sustain significant losses. For that reason, it always relies on an 'air strike' policy, which does not allow for the establishment of a *pax americana*. It is said that the United States is a superpower without a sword. This is, to a great extent, true.[12] After the dramatic lesson of Vietnam, the United States is now very careful not to intervene militarily with ground troops. It has only limited strategic military objectives. The new doctrine of the Pentagon is to prepare the military capacity of the United States to fight in only two regional conflicts at a time, in any part of the world, and even that only in extreme cases, where its vital national interests are threatened.[13]

Today the nuclear arsenals of the United States, Russia, Britain, France and China have lost their strategic significance; this is because the socio-political turbulence in the different regions of the world cannot be resolved by nuclear dissuasion. The strategic situation of the world is complex because it has emerged from an era of strategic stability anchored in the Cold War, and entered upon an unstable process of global disorder. The same bipolarity that for forty-five years affected the conflicts of the so-called Third World, also served to control them. The US–Soviet rivalry was a factor that regulated ethnic violence and historic rivalries. The local conflicts were selected, limited, controlled, given 'low intensity'. The Cold War constituted a kind of violence controlled by the superpowers in order to avoid a direct confrontation.

With the end of the Cold War, the strategic dyke that the two superpowers had built so as to contain violence in the world was breached. An avalanche of political disintegration, insurrections, civil wars, ethnic or religious conflicts, massive violations of human rights, genocide, waves of refugees and displaced persons was the result. The unresolved historical conflicts and rivalries raged out of control and acquired their own dynamic, fed by myths, pocket ideologies, tribal nationalisms and messianic fundamentalisms. The Balkans and the Caucasus blew apart. Famine, violence and genocide broke out in Liberia, Somalia, the Sudan, Angola, Mozambique, Rwanda, Burundi and Congo (ex-Zaire). Civil war began in Afghanistan and Sri Lanka; guerrillas were active in Colombia and terror rampant in Peru; fundamentalist movements grew in Egypt, Algeria, Pakistan and India, and new guerrillas emerged in Mexico.

The situation in many quasi nation-states, submerged in violence and on the brink of collapse, obliged the United Nations, which had been designed to confront international conflicts *between* states, to become involved in difficult *internal* conflicts and civil wars, and to increase its peacekeeping operations. Nearly sixty thousand blue-helmeted UN troops were mobilised, to the tune of 4 billion dollars annually, in order to contain this tidal wave of violence. Nowadays, the UN is discredited and practically impotent to contain the depredation of nations that are collapsing in bitter domestic fighting.

When the Cold War ended, politicians, diplomats, economists and experts in international relations never imagined that the world situation would evolve into a sort of modern barbarism. On the contrary, it was thought that, after the collapse of Communism and the success of collective security in the Gulf War, we were poised on the threshold of a new world order based on capitalist democracy and global prosperity. It was in this context that the Secretary General of the UN, Boutros Boutros-Ghali, held the candid belief that he could become a kind of general manager of the new world order. After all, he was the first Secretary General who was not going to have to face the paralysis of the UN caused by the cold war rivalries. Convinced of his new mission, he concentrated the UN's activities on peacekeeping operations, confident that the great powers would support him and use the Security Council to put the world in order.[14]

However, the great powers' commitment in these operations was mediocre: they merely created a false and precarious climate of security which permitted humanitarian aid to reach the victims of civil war without guaranteeing their physical safety, and without punishing the aggressors or detaining those accused of crimes against humanity. The Security Council organised strange hybrid interventions, combining humanitarian assistance with peacekeeping operations, which soon proved to be ineffective since they neither punished the aggressors, nor saved the victims, nor enforced the peace. The most blatant examples of these fiascos were the former Yugoslavia, Somalia, Rwanda and Congo (ex-Zaire).

The discrepancy between Boutros Boutros-Ghali's pretensions and the great powers' reluctance to use their swords for the purposes of putting order in the world ended with the Secretary General's not being re-elected, due to a United States veto. This distinguished civil servant apparently was unaware that, nowadays, the so-called great powers have enormous difficulties in intervening militarily to set the world in order. This is not the result of lack of political ambition, but is rather the consequence of a problem of civilisation. Their consumer societies, based on the principle of instant gratification, are unwilling to accept sacrifices to correct evils in poor and distant regions of the planet. The politicians of the great powers find it nearly impossible to sell their fellow citizens the idea that it is essential to participate in the 'just wars' of the United Nations. Their electorates refuse to sacrifice the lives of their sons and pay more taxes in order to establish a new world order. No consumer society wants to take on the human or the economic costs implied in 'peace making'. The mere idea of television pictures of their soldiers returning in bodybags terrifies governments, because of the potential backlash from voters. In consequence the governments of the great powers have adopted the policy of military interventions with 'no casualties' as their norm, and have therefore been extremely prudent in embarking on UN peacekeeping missions. Their political practice in recent years since then has thus been to safeguard the national electorate by abandoning world order.

Today, the great powers do not function and, as a result, neither do the United Nations' peacekeeping operations. The great powers' answer to the disorder in the world is always a combination of extreme prudence and cynicism, disguising their lack of power. This is the main cause of the malfunctioning of the world organisation, a cause that these powers either ignore or try to ignore, instead criticising the UN as though it were in itself a great world power rather than the reflection of the policies of swordless powers like the United States, Russia, France, Britain, China, Japan and Germany.

A nation-state can only be called a great power if it exercises a power policy – in other words, if it is willing to use force and to suffer many casualties, if it refuses to be humiliated and if it inspires respect. To lose a thousand or more men in an imperialist policy, in colonial wars or in punitive expeditions was routine when the great powers behaved as such in Asia, Africa, Central America and the Caribbean. In Somalia in the 1990s, by contrast, the death of a handful of marines triggered the evacuation of US forces. The United States considered that Yugoslavia's disintegration was a European problem. In like manner, France and Britain refused to send troops to Bosnia and submitted to blackmail and humiliations from the local forces. In the end NATO intervened, but only after crimes against humanity had already been perpetrated. NATO then organised air strikes, which failed to solve the problem of Kosovo.

For their part, the United States and Europe have maintained a prudent distance from the genocide in Africa, in Sierra Leone, Liberia, Rwanda, Burundi and Congo (ex-Zaire). It could be argued that none of these great powers has interests in Africa, but precisely their lack of interest in filling a void in the region, using the opportunity presented to protect human rights, is proof that they have lost the instincts and behaviour of the great powers.

This loss of a bellicose spirit in the developed societies is not a negative historical phenomenon, even though its origins may be judged egotistical. Loving to enjoy life is, certainly, a vaccine against fanaticism, and it also fosters a critical spirit against the *causes patriotiques nobles* that drove millions of young Europeans, Americans, Russians and Japanese to slaughter in the two world wars of the twentieth century. Youths in prosperous societies are every day less inclined to follow bellicose dictates, and that is good. However, their immersion in an instant gratification society, indifferent to the 'new barbarians' in the South, is creating a moral crisis in the great powers. This is increasingly evident in their spiritual void, a sentiment of non-commitment to humanity, and in the diversion of their youthful energy into earning easy money or into violence and drugs.

Today, in short, the nuclear arms of the most powerful nation-states are of no use against the domestic infernos caused by genocide and outpourings of refugees, whilst the use of conventional military forces is seriously curtailed by the 'no-casualty' intervention policy. If like the United States, Britain, France, Japan or Germany, they have prosperous societies, the material gratification that these provide dissuades from military intervention. If, on the other hand, a great power lacks a prosperous society that can offer its citizens material gratification, it collapses, as did the Soviet Union. Impotence, caused by either an excess or a lack of prosperity, is a new international phenomenon, not experienced in the two hundred years since imperialist policies were first applied.

The fading-away of the policies of power is the clearest proof of the twilight of the nation-state, including its most powerful representatives.

Surrogate power

As a consequence of the decline of the nation-state, the search for development is beginning to go beyond the confines of that political structure. This myth, which was formerly sought by strengthening the state's role using economic and cultural protectionism, is now pursued through the competitive integration of the national economy in the global economy.[15] This is causing many cities and regions in the wealthy countries to begin to internationalise their activities outside the nation-state, so that they become gradually more autonomous. Today the global economy is defining the real human frontiers, which are not necessarily the territorial boundaries of the nation-states but are instead transnational economic zones resulting from the penetration of sovereignties. In the rich economies, a kind of global economic tribalism is proliferating: Scots, Catalonians, Basques, Lombards, Walloons, Alsatians, Bavarians, Quebecois, or Californians are increasingly vociferous in their demands for autonomy, so as to integrate their regions or cities directly into the global economy.

In Europe, this malfunction of the nation-state is resuscitating the city-state. Cities like Lyon, Milan, Stuttgart and Barcelona have managed to escape, by means of a co-operation pact, from the excessive control of Paris, Rome, Berlin and Madrid. In the Benelux countries, Antwerp and Rotterdam are together becoming one of the most dynamic autonomous economic centres of the continent. Liège, Maastricht and Aachen have revived their medieval community, forming a triangle of prosperity that crosses the Dutch, Belgian and German borders. The Nordic cities are resuscitating the Hanseatic League with the Baltic cities. Everything seems to indicate that the city-state will be an important actor on the economic scene, as well as a factor of human identity in twenty-first century Europe. This trend could spread across the globe, in view of the fact that, in this new century, many cities in the world will have populations of 10 to 20 million inhabitants.

China, a totalitarian state, had to concede a series of auto-nomous economic zones on its coast. These have joined the trans-national sector, and are virtually disconnected from the poverty of deepest China. Guangdong Province, with Hong Kong as its capital, Shenzhen, Zhuhai, Amoy and Guangzhou, enclaves with considerable autonomy, are more connected to the global economy than they are to Beijing. This situation has created a loss of central control that an engineer of totalitarianism, such as Mao, never could have imagined.

As cities and regions gradually gain a position as actors on the international stage, this is giving rise to a metropolitan micro-diplomacy that undermines sovereignties even further. This new phenomenon constitutes a kind of prosperous and positive disinte-gration of the nation-state's power. Conversely, in many poor quasi nation-states, where the most explosive urban population growth is occurring, state centralism, as it continues to avoid granting auto-nomy to its gigantic, poverty-stricken cities, may bring about the disintegration into poverty of the nation-state. Without autonomy, these enormous human settlements will become environmental bombs and domestic infernos as a result of pollution, unemploy-ment and delinquency.[16]

In our time, what happens to the poor cities will increasingly determine the fate of the underdeveloped countries. Between 1950 and 1995, the number of cities with more than 1 million inhabitants has increased sixfold in those countries. The UN estimates that the poor cities will continue to grow explosively and that more than 5 billion persons, that is 70 per cent of the human race, will live in poor cities in the year 2025.

Faced with this prospect, some experts think that, in order not to be left out, the underdeveloped countries should organise themselves in smaller and more efficient territorial subunits in the area of a large city, connected autonomously to the global economy. These new territorial subunits would have only 5–10 million inhabitants. That is, small enough communities so that their citizens would share common interests and adequate infrastructure could be built, with a view to integrating efficiently with the global economy. Within this picture, the central government of underdeveloped states with very large territories, suffering social disintegration and with problems of population explosion and economic viability, should gradually concede autonomy to these subunits through a democratic process, retaining functions only in the areas of defence and diplomatic relations.[17]

According to this globalist approach, it is a matter of promoting positive national disintegration by means of micro-territorial economic units (cities with small regions) connected directly to the global economy through transnational corporations. A sort of global network would be formed, made up of hundreds of new Hong Kongs and Singapores. These would make the world market work more efficiently and provide well-being, in contrast to what is happening today in the majority of underdeveloped countries, whose centralism, enormous territories and huge populations render them too large to be prosperous in a changing and highly competitive globalised world.

The decline of the nation-state is visible not only in the loss of power of the great powers and in the search of their own regions and cities for well-being via globalisation. At the beginning of the twenty-first century, sovereignty of the most powerful Leviathans has been undermined by transnational corporations. As a result they cannot control their national economies and are incapable of building a new world order and establishing an efficient international system that protects the planet from environmental deterioration.[18]

Even the most powerful and modern industrialised nation-states are losing more and more of their control over their national economy to a globalisation that is driven by the transnational corporations born out of their own capitalism. Meanwhile, the few underdeveloped quasi nation-states that have industrial sectors with some global competitive advantages have to associate with the transnationals as the only channel for continuing to attract investments and technology and thus maintain these advantages. The worst situation is that of the quasi nation-states that are mainly dependent on primary exports with low technological content, that have an overflowing urban population and that do not receive a critical mass of productive transnational investment.

All the nation-states are becoming promoters of transnational investment. Since 1980, all without exception have changed their economic policies, liberalising, deregulating and privatising their economies. Thus they create conditions that permit the transnational corporations to enter their markets, which is tantamount to abandoning their national capitalism to the mercy of global competition. The nation-state is withdrawing from the economic and financial domain and giving way to transnational globalisation. It is becoming more an administrative than a sovereign territorial entity, a kind of 'surrogate for transnational capitalism'. Now its main role is that of an efficient manager, with a mission to liberalise and deregulate, to supply good infrastructure, to flexibilise employment and to strengthen public security, in order to foster a positive investment climate for transnational enterprises.

This surrogate transnational power may be able to create prosperity, but the system will surely undermine both national capitalism and, with that, democracy, the two foundations of the modern nation-state. As a great part of national capitalism is absorbed by global capitalism, the citizens will be unable to exercise democratic control over the economic policy of their countries, because unknown persons who were not elected by them will take many of the decisions outside their borders. The discussion today is not about a viable alternative to capitalism, because that does not exist. What is being debated is whether the national capitalist economies should virtually disappear within a global capitalism that is beyond democratic control because it lacks territoriality.

In spite of the risks, the nation-states, like fading old aristocrats, are trying to fit in as surrogates of an emergent world aristocracy led by the most powerful transnational corporations. If this surrogate system works, as is the case in the industrialised nations and in a handful of newly industrialised countries, production is modernised and it is possible to compete in the global market. Of course, the price to be paid is a loss of national power and of democratic control over the economy. If it does not work, as has been the case up until now with all too many poor countries deprived of productive transnational investments, these countries' exclusion from the modern global economy will increase and, with it, their national poverty and the probability of their imploding into ungovernable entities.

CHAPTER 2

GLOBAL EMPOWERMENT AND NATIONAL IMPOVERISHMENT

The new global aristocracy

World power has always been a game of geopolitical balance among a diminishing aristocracy of great powers. From the nineteenth century to the First World War, the players were Britain, Germany, France, Russia and Austro-Hungary. Between the end of that war and the Second World War, the game continued with the United States, Britain, France, Germany, Japan, Italy and the Soviet Union as participants. Later, the game was drastically reduced, to just two superpowers, the United States and the USSR.

The game ended with the end of the Cold War. The United States, the only superpower, cannot play the global equilibrium game, since it has no military rival. Even so, it lacks sufficient military and economic capacity to impose order unilaterally in the world and control the global economy. As a consequence, in the twenty-first century, world power will no longer be the result of a game of geopolitical balance among great powers. Power will not be gauged by the number of divisions, planes, fleets and nuclear missiles belonging to any given country; rather it will depend on its capacity for technological innovation, in order to compete and do business on a planet-wide scale. The new game for world power will no longer be geopolitical but instead geo-economic; its main actors will be not the former world aristocracy of great powers but a new aristocracy, the transnational corporations. These are beginning to dominate the world, penetrating all the national sovereignties with their merchandise, services, capital, technology, telecommunications, credit cards and patterns of consumption.

Today 38,000 transnational corporations and their branches conduct two thirds of the world's trade, and the combined sales of the eighty-six most powerful enterprises are larger than the exports of nearly all the nation-states that constitute the present international community. Only the exports of the nine most industrialised powers – the United States, Germany, Japan, France, Britain, Italy, Canada, Holland and Belgium – exceed the sales of Shell, Exxon, General Motors, Toyota, Ford, Mitsubishi, Mitsui, Nissho Iwai, Sumimoto, Itoch Maruben and Hitachi, the ten most powerful transnational corporations.[1]

The world power that was the exclusive province of the old aristocracy of great, industrialised nations is now beginning to belong to this new non-state international aristocracy. In the last quarter of the twentieth century the transnational corporations have proliferated: from 7,000 enterprises to nearly 38,000, with 250,000 subsidiaries, spreading consumption patterns and a similar lifestyle all over the world. The new global aristocracy decides worldwide where, what, how and for whom to produce.[2] Today, the destiny of many national economies and cultures is being decided not in government offices or parliaments, but in the international financial markets of New York, Chicago, London, Singapore, Hong Kong, Tokyo, Frankfurt or Paris, and in the boardrooms of the transnational corporations.

International trade today is virtually a subproduct of the investments, alliances and agreements among transnational corporations. At present, nearly 70 per cent of world trade takes place among those enterprises. These new economic relations have nothing to do with the famous comparative advantages of the different countries cited in the neoclassical textbooks of economics. In their alliances with suppliers, in their licensing agreements, in their franchise contracts and in negotiating their global strategy, these enterprises are not really exposed to the policies of the free market so in vogue in the political and academic discourse of the end of the twentieth century. Present-day globalisation is the result not so much of free global competition among nations, but of a network of agreements and productive and financial activities among the transnational corporations. A large part of international trade and finance is still registered nationally by countries. This is the case not because the countries are in fact carrying it out, but simply because the transnational corporations' goods and services cross their borders.

These gigantic corporations, which used to be viewed with such fear as manifestations of imperialism, are now considered the embodiment of prosperity and modernity. All countries try to attract the investment and technology of the transnational enterprises, in order to increase the comparative advantages of their national economies and to gain markets. Sony purchases part of Hollywood, with the blessings of the North Americans, while Microsoft is welcomed by the Japanese and the Europeans. Everyone wants Mercedes to invest, and no country can feel truly modern without CNN. The truth is that it is practically impossible for a national economy to have a globally competitive export sector without being associated with some transnational enterprise. There is no doubt that, if Singapore, Hong Kong, Switzerland, Germany, South Korea and the United States itself have great competitive segments in their economies, they are the result of transnational operations within their borders.

The underdeveloped countries, with their lack of national capitalism, huge rates of unemployment, high demographic growth, and raw material exports at unprofitable prices, have no choice but to seek productive transnational investment. Only in that way can they hope to reduce somewhat their unemployment, increase the technological level of their production, and develop new exports with comparative advantages. For these reasons, there is at present a dearth of transnational investments in all the countries of the world. It is not an easy task to become the country chosen for investment by the much-sought-after and spoiled aristocrats of the global economy.

The transnational corporations are very cautious and selective in their investments. They are interested only in the national factors that will produce the highest profits without running any great risk. They are particularly attracted by the technical capacity and productivity of the workers, by the opportunities for sub-contracting to national companies with technological capabilities, by good infrastructure, by the size of the domestic market, by the legal security and the political stability of a country.

Until the present, these conditions have been found, according to the transnationals, only in twenty-four industrialised countries, above all, the United States, Japan, and the European Union, which in transnational terminology are called the TRIAD. That is where they place 70 per cent of their productive investments. As a second priority, they prefer the Asia–Pacific region, with countries like China, Singapore, South Korea and Taiwan. The third priority, with much smaller investments, includes certain Latin American and former Communist countries, such as Argentina, Brazil, Chile, Colombia, Venezuela, Russia, Poland, Hungary and the Czech Republic.[3] The remaining countries of Latin America, Asia and Africa receive very little productive transnational invest-ment, despite the market reforms and the incentives that they offer, because they are considered to lack the prerequisite condi-tions needed to make the profits these corporations hope to obtain.[4]

By concentrating their activities in the TRIAD, the trans-national corporations exert a strong influence on the political sector of the industrialised powers, which they apply through battalions of lobbyists and by dint of economic contributions. That is the way they woo the economic diplomacy of the United States, the European Union and Japan to promote their global interests. Today, on the international scene, the former world aristocracy of the industrialised nations, represented by the Group of Seven wealthiest countries in the world, frequently practises surrogate diplomacy in favour of the transnational companies, applying pressure to gain access to foreign markets for their products, services and capital.

The most powerful transnationals have thus succeeded in convincing the diplomats of the great industrial powers to capture the major international economic agendas in favour of their global interests. In the 1980s, these were the powers that turned the IMF into a collecting agency for the foreign debt of Latin America, thus guaranteeing the interest payments to the international creditor banks. They also managed to use the Uruguay Round of the GATT to gain access to nearly all of the national markets for transnational services and to develop a stringent international system of intellectual property rights protecting the technologies of the transnational enterprises. In the recent Asian financial crisis, the IMF was converted by the Group of Seven into a gigantic provider of liquidity, so that the bankrupt Asian countries could pay the bad loans made by many transnational banks and investors.

Nothing proves more clearly the world power of this new non-state transnational aristocracy, and the surrogate role of the great industrial powers in its favour, than the draft of the Multilateral Agreement on Investment (MAI) of the OECD (the twenty-four most industrialised countries). In the end, the draft was not adopted because it went so far as to attribute rights only to transnational corporations and only duties to the Nation-States. It even included the right for transnational investors to appeal to the courts and demand compensation from governments for loss of earnings. This new global system for transnational investments is an attempt to consolidate definitively the hegemony that transnational capitalism already exercises over the states and the national capitalists.

The constant pressure exerted by the Group of the Seven most industrialised countries for continued liberalisation of manufacturing, services and capital markets is aimed at converting the highest-income groups in the poor, most populous countries into clients of the transnational corporations. From this point of view, the opening of the Indian and the Chinese markets is of major importance, since this move captures at least 400 million high-income clients, or a market equivalent to that of the United States or Europe.

The economic surrogate diplomacy of the old industrialised state aristocracy in favour of the new transnational aristocracy is not restricted merely to gaining access to the markets of India and China. It also tries to gain some millions more clients in the Latin American, Asian and African countries. This is the logic behind present-day globalisation. The whole process would not give reason for concern if it were creating abundant employment and if the poor countries' exports were growing at the same pace as the transnationals' sales to them. In fact, the contrary is occurring. Today, nearly all the underdeveloped countries have become large importers of transnational goods, and many of them are suffering serious external deficits. However, this external imbalance is not discussed at the new World Trade Organisation since, for the new transnational aristocrats and their surrogate governments, the purpose of free trade is to find clients and not to help countries to develop.

In spite of the surrogate relationship between the great industrialised countries and the most powerful transnational corporations, the latter do not feel any identification with the national interests of the former. Coca-Cola, Nestlé, Nike or Nissan do not identify themselves with the US, Swiss or Japanese national interests, since their products are manufactured in many countries and are sold throughout the world. In pursuing their global objectives, many transnational enterprises have even relocated activities from the very capitalist national economies where they had originated, thereby creating unemployment in these economies.

The operations of a large majority of the transnational corporations are distributed all over the world. They are allied with other transnational enterprises, and their executives include a wide variety of nationalities. If their headquarters were not located in one of the countries of the TRIAD, no one would really know what their nationality was. Nowadays it can be said that the largest transnational corporations really have no nationality; instead they have a national origin and broad global interests.

The strangest fact about the emergence of this new non-state aristocracy is that the transnational enterprises enjoy increasing world power, whereas, paradoxically, they assume no international responsibilities. In contrast to the old aristocracy of the great powers, who tried to balance their national ambitions against their international responsibilities, such as the protection of human rights or of the environment, the executives of the transnational corporations do not want to establish any link between their global negotiations and the problems that these often cause. To mention only a few of these problems, they frequently generate unemployment, cause environmental damage or depend on complicity with oppressive regimes. It would be very surprising if the new transnational barons were to consider China's record on human rights in their decisions about investing there, or were to decide not to use a new labour-saving technology in Nigeria in view of that country's unemployment problem. It would be almost utopian to think that they would increase their oil production costs and transport costs in order to avoid the present risks of marine pollution. For the executives of the transnationals, global social, economic and environmental problems pertain exclusively to the competence of governments, and the latter should resolve them without interfering in the transnationals' markets.

The only responsibility that the transnational corporations recognise is to their stockholders. However, the speed with which the ownership of their shares is interchanged on the global market makes it nearly impossible to identify the owners. No one knows who they are. The worldwide power of transnationalism today is increasingly anonymous. So much so, that unknown persons who were not elected decide the value of a country's currency, the price of its raw materials, the cost of credit, and the prices of energy or food. They thus affect the fate of millions of people in many nations. In consequence, the very essence of democracy begins to erode, because the citizens have no way to influence the conduct of their own country's economy. They no longer feel that their governments represent them, as governments lack power to change the situation and even act like surrogates for the transnationals' interests.

Gradually, a kind of anonymous global economic and financial dictatorship is being established as a result of the transnationals' lack of responsibility in the face of unemployment, financial speculation, currency fluctuations and environmental disasters. As globalisation of the market increases, democratic national control of the economy diminishes and, along with it, the feeling of belonging to a nation or a community. In this way, resignation and social anomie arise, but so do frustrations and violence. In the end, the governments have turned over their domestic and international markets to the transnationals without demanding in reciprocity a joint responsibility for solving the problems that the globalisation of the economy is creating.

The only way to force the new international non-state aristocracy to take on responsibilities is simply to grant them international representation, with full membership in the international economic, financial and environmental organisations. The largest and most powerful transnationals dominate almost entirely the production of and world trade in energy, metals, chemicals, food, biotechnology, automobiles, aerospace, electronics, computing, telecommunications, transports, media, finance and banking. Therefore, they cannot continue to stand aloof from the world's economic, financial and environmental problems. Today, the member states of the UN, the IMF, the World Bank or the WTO are still looking at themselves in a sort of twentieth-century magic mirror. This image constantly deceives them, telling them that they are the stars of the contemporary international stage when that is no longer the case. Maintaining an exclusive representation of nation-states in the World Bank, the IMF or the World Trade Organisation is totally absurd in view of the fact that the GNP and the exports of nearly half of the member countries of these organisations are less than the sales of the 100 most powerful transnational enterprises.

With the exception of the grand show at the Davos Forum, there are no permanent and effective international stages where the governments of countries can meet with the representatives of the transnational corporations to negotiate about investments, environmental protection, and the use and acquisition of technology. The underdeveloped nation-states and their national business leaders would no doubt be very interested in having a place where they could deal with these and other questions with the most important actors in the global economy. Until now, the UN programmes for environmental co-operation and development, as well as the programmes of the World Bank, have been negotiated without any participation by the transnational corporations.[5]

The international organisations cannot continue to be mere diplomatic forums, places for speeches, discussions and negotiations between government representatives that have no real power to change global economic and environmental trends. The concept of international representation and co-responsibility must be extended to include transnational enterprises and representatives of civil society. Only then can the international organisations truly reflect the real world. How can the environment be protected if there are not even consultations with the company that is causing the pollution, or with the company that invented the technology for pollution control? How can the modernisation of sectors of the underdeveloped economies be programmed without contacting the possible investors? How is the volatility of the world financial market to be managed if the new barons of real world power are not made co-responsible?

Today we are witnessing the emergence of a new and powerful non-state world aristocracy alongside the decline of the old aristocracy of nation-states. Between the two processes, a power vacuum is forming. Governments do not have the power to resolve the world's problems on their own, and the leading transnational enterprises do not want to assume any responsibility for these problems. The old world aristocracy and the new are behaving as though they were not living on the same planet, as though they or their descendants will be untouched by the tensions and the violence that are already rising as a result of globalisation, as though they will be untouched too by the fact that no one is trying to deal efficiently and seriously with social exclusion and growing environmental degradation.

The supranational clergy

In the course of the last twenty years, not only has a non-state world aristocracy arisen, with more global power than the nation-states, but a powerful international bureaucracy has also been created. This body was not elected by the people, yet it establishes the rules of the economic game for the majority of the world's population. Today the IMF and the World Bank have acquired supranational powers to dictate and supervise the economic policies of any developing country, affecting for good or evil the daily life of every one of its citizens, without being accountable to anyone.

In the face of this undemocratic supranational power, it is often argued that it is not the IMF or the World Bank that impose such policies on the citizens; rather it is their governments that accept them. This may be the case in theory, but in practice governments have no alternative to the rigorous policies of the IMF and the World Bank. If a government rejects them, it becomes a world pariah, with no access to international credit. The African leaders have an expression to describe this inextricable situation. They call it TINA, meaning, 'there is no alternative'.

This supranational power of the IMF and the World Bank was instituted, without the slightest international legal foundation, in the 1980s. It was a response to the pressure brought to bear by the transnational banks and the creditor governments to obtain guarantees for the payment of the foreign debt. The IMF and then the World Bank were converted into strict supervisors of an adjustment designed to oblige the debtor countries to reorganise their economies and to pay their creditors.

The supranational monitoring, by the IMF and the World Bank, of the national economic policies of the Latin American, Asian and African countries is a new phenomenon in contemporary international relations. It clearly reflects the erosion of national sovereignty and the emergence of a supranational power over an entire category of quasi nation-states that call themselves sovereign. In point of fact, under the supervision of the IMF and the World Bank, the so-called developing countries, from the least poor, like Venezuela, to the most poverty-stricken, like Mozambique, have lost democratic control of their national economic and financial policies. This supranationality is continuing to spread, and threatens even to overrun political aspects that were formerly the exclusive province of the sovereign state, such as state objectives, governance, corruption or military expenditures.

Today all the governments of the underdeveloped countries are subjected to public policies designed by an anonymous international technocracy that has not been elected by their citizens. Since they have no alternative, they are careful not to submit these supranational policies, which affect the daily life of their citizens, to their parliaments; much less do they seek the approval of a popular vote. Many newly elected governments immediately renounce the promises of their election campaigns and apply instructions from the IMF and the World Bank. Many elected governments in Latin America, Asia and Africa that call themselves democratic are rapidly being transformed, by supranational economic policies beyond their citizens' control, into very low-intensity democracies.

Given their nearly exclusive dedication to the underdeveloped countries – countries that have no influence on the functioning of the world economy – the IMF and the World Bank have neglected many important problems inherent in the current globalisation process. They are not seeking solutions to control the highly volatile nature of the global capital speculation market, nor are they concerned with establishing mechanisms to include the political participation of the transnational corporations, the main actors in the global economy.

Such virtually exclusive dedication to the underdeveloped countries, along with the submissiveness of their governments, has transformed the IMF and the World Bank into a powerful and illuminated supranational high clergy. They are devoted to preaching, with great fervour and conviction, a 'single economic creed' for the salvation of all the underdeveloped countries from the nightmare of poverty and their conversion into emerging capitalist economies.

This single creed is fundamentally contained in what John Williamson, of the Institute of International Economy, called in 1990 the Washington Consensus. This comprised a series of principles of economic policy that emerged from the continuous consultation of the Congress and government of the United States, the IMF and the World Bank with bankers, transnational executives, politicians and finance ministers. The central message of the creed is: the free market should regulate all economic activity; the states should intervene to maintain fiscal discipline, attain a stable rate of exchange, liberalise, deregulate, privatise the economy, as well as to make employment flexible, as the only way to gain access to credit and attract foreign investments.[6]

The success of the missionary task of the World Bank and the IMF, which have since been joined by the WTO, is unquestionable. This supranational high clergy has managed to convert nearly all the countries of the underdeveloped world to the single creed of the Washington Consensus. Today there is a worldwide consensus to the effect that there is no alternative to these principles of economic policy. The majority of the leaders of the poor countries believe that severe adjustment and the violent shock of opening up their national markets are the necessary penitence that will enable them to absolve the sin of poverty and to enter the holy land of prosperity, like new Singapores, Koreas or Taiwans. Other leaders, however, are more sceptical, but think resignedly that these economic policies at least will serve to keep their countries from becoming international pariahs. The IMF, the World Bank and the WTO have thus achieved simultaneous, universal enforcement of one single faith on the majority of the world's population, a feat that no religion has ever attained. Their economic credo reigns as the only option in the world, and the supranational clergy is today powerful and feared.

Nevertheless, this success of the supranational high clergy is limited to its preaching and conversion work; it has not managed to obtain tangible results in terms of salvation from the nightmare of poverty. Adjustment and the violent shock of opening markets have not yielded the desired success. While many poor countries stabilised their economies, lowered inflation and achieved modest economic growth, this only helped them to import more global products from the transnational aristocracy, without diminishing their poverty. After more than twelve years of applying adjustments and reforming their markets, the great majority of Latin American, Asian and African countries are still trapped in the purgatory of neoliberal reforms, and have not managed to break free from the sins of unemployment and poverty.

Their raw material exports do not fetch profitable prices, their debts continue to be a heavy burden, their imports increase, and productive transnational investments are not forthcoming; nor have new companies been formed in their territories to export products with more technological content, to make them competitive in the global economy. They carry through the programmes of liberalisation, deregulation and privatisation, but genuine market economies do not emerge because a vast proportion of the population live below the poverty line. In these circumstances, the much-needed broad middle classes do not arise. Neither do modern capitalist democracies, in which the law and institutions are upheld and a culture based on competition replaces the prevalence of favours and corruption.

The most glaring example of the failure of the supranational high clergy has been the collapse into bankruptcy of the emerging economies of Asia, and the IMF intervention – violating the principles of the free market – with a multimillion-dollar financial package designed to save investors and transnational lenders from insolvency in Thailand, Indonesia and South Korea. Intervening to save capitalists who make bad investments flies in the face of the single creed that this supranational high clergy has been preaching, since free economies have a very effective method of punishing reckless investors. It is called bankruptcy. In this case, those who were punished by the IMF were the ordinary people and the middle classes of South Korea, Thailand and Indonesia. They had to withstand a punishing adjustment programme in order to pay this ransom.

The concrete consequences of IMF and World Bank policies in the underdeveloped countries have awakened strong criticism around the world, from outstanding economists, business leaders and religious sectors.[7] The criticisms are based on twelve years of observation of similar neoliberal policies in the United States and Europe that have only increased social inequalities and unemployment. In the United States today, social inequalities are greater than before the Second World War, while in Europe unemployment has increased to unprecedented levels. The critics of the supranational high clergy wonder how such policies, which have created social exclusion and unemployment in the most prosperous countries, will be able to create no less than 1 billion new jobs and rescue from poverty 4 billion persons in Latin America, Asia and Africa, whose incomes are less than two dollars a day. According to the Reports on Human Development of the United Nations Development Programme (UNDP), after twelve years of adjustments and reforms, the income gaps between rich and poor people are widening continually – in the wealthy as well as in the poor countries. The World Bank itself is now concerned about world poverty and admits that the underdeveloped nations are a very long way from becoming newly industrialised countries.[8]

It is important to point out that all the criticisms from American and European businessmen, economists and religious leaders of the policies of the IMF, World Bank and WTO are not directed against capitalism. On the contrary, all these critics recognise that the market economy is the most efficient economic system or, at least, the lesser evil of those that have been tried to date. Their argument is, rather, that the neoliberal shocks that are applied to the underdeveloped countries in order to turn them into modern capitalist economies overnight are a new experiment. They were never applied to the capitalist development of Europe, the United States, or Japan, or to the Asian NICs. All these countries grew as modern capitalist economies by protecting and helping their national capitalism. The United States itself has a most convincing example, in its defence procurement policy, of state industrial policy in favour of its national enterprises. For more than half a century, the defence procurement policy of the United States has acted as a true state intervention policy, granting subsidies to its industries and supporting its national technological development.

In spite of this historic experience, the supranational high clergy obstinately continues to believe in the infallibility of its single creed, derived from the principles of the Washington Consensus. This attitude worries many academics, because it is based more on ideology than on reality. In fact, many of the principles of the single creed are founded on neoclassical theoretical economic postulates that do not apply in reality as scientific law. The first of these is the theory of the 'comparative advantages' of the under-developed countries. According to this theory, a hyper-free market is required so that the advantages that the countries of the underdeveloped world have in terms of cheap labour and natural resources can attract the necessary transnational investments. In fact, today, the comparative advantages of the underdeveloped countries are no longer determining factors for attracting foreign investment. The transnationals do not place their investments according to the supply of natural resources or cheap labour, but with a view to the hefty profits to be gained from the high rate of productivity provided by a highly skilled workforce and local enterprises with technological capabilities. These factors do not exist in the majority of the so-called developing countries. It is precisely for these reasons that the transnationals mainly invest in the TRIAD.[9]

The second theoretical postulate is reflected in the concept of 'perfect competition' of the Victorian economists and their neo-classical present-day followers. According to this theory, the most absolute economic freedom is required for demand to equal supply in the market. If they do, all the economy's resources will be totally absorbed and all the economic participants will enjoy the highest degree of prosperity. In this theoretical world of competi-tive equilibrium, the state should not intervene in the economy to balance supply and demand. They are self-adjusting and any intervention could be damaging for the actors, whether enterprises or persons. As a consequence, the state should be as small as possible, and fade away to the greatest extent feasible. This perfect equilibrium has never been attained in any national reality, no matter how many mathematical formulas are used to explain it in the modern textbooks on neo-classical economics.[10] In practice, the only countries that have managed to escape from underdevelopment have been countries in which the state, by becoming the entrepreneur, has given support to the efficient capitalists with potential comparative advantages for exporting, as happened in South Korea, Taiwan and Singapore.

Perhaps all this theoretical positing is not quite as utopian as the premises of the world proletarian revolution. Ironically, the supranational high clergy coincides with Marx in one central point of doctrine. Both want solely the material prosperity of the world and the disappearance of the state: Marx, through the transnationalisation of the proletariat; and the supranational clergy, by means of the transnationalisation of capital. Against his wishes, Marx achieved the contrary. His doctrine inspired totalitarian states. Will something similar happen to the supranational high clergy? Instead of global capitalist prosperity, will they produce a Darwinian world, plagued by unemployment, social exclusion, violence and environmental deterioration? Global indicators would suggest that the world is headed in that direction.

The international third estate

In the face of the great emerging power of the new transnational aristocracy and the policies of the supranational high clergy embodied by the IMF and the World Bank, the power of the so-called developing countries is marginal. Even the most powerful of that group, such as Brazil and India, are unable to influence the rules of the international economic game.

The feeble economic power of all those countries is revealed by the fact that each of the 100 most powerful transnational corporations sells more than the export totals of each one of the 120 so-called developing countries. What is more, the 23 most powerful transnationals sell more than the exports of the developing countries with the largest economies, such as India, Brazil, Mexico, Argentina and Indonesia.[11]

In addition, with the end of the Cold War, the countries that make up the Third World have lost their relative strategic power to bargain for modest concessions from the industrialised powers and to orient their economic policies without totally accepting the supervision of the IMF and the World Bank. Today, all these countries blindly obey the single creed based on the Washington Consensus. They have all liberalised, deregulated, and privatised their economies. Now, at the mercy of the transnational corporations, they await the investments aimed at their technological modernisation and the creation of enterprises capable of competing in the global economy.

The Non-Aligned Movement and the Group of 77 continue to perform their diplomatic ritual in the United Nations, although their negotiating position is virtually extinct. The Third World as a political entity has ceased to exist. Perhaps, most probably, it never did exist. It was just a diplomatic stand that appeared to be solid and to possess a negotiating power that it never really had. In point of fact, even during their halcyon days as negotiators during the petroleum crisis of the 1970s, the Group of 77 and the Non-Aligned Movement failed to extract any concessions from the aristocracy of the industrialised states in order to revalue the prices of raw materials and to lay the foundations for a new international economic order. The maximum that these countries obtained was that important conferences were convened, such as those of UNCTAD and the North–South Summit dialogue. Despite these meetings, raw material prices not only were not re-evaluated, but, on the contrary, they fell. The sole exception was petroleum, whose price rise was an even greater blow to the Third World countries than to the industrialised nations, obliging them to borrow and precipitating the debt crisis.

The Uruguay Round of multilateral negotiations (1986–94) was the clearest proof to date of the so-called developing countries' lack of negotiating power, and a definite demonstration of the growing power of the new world transnational aristocracy to obtain through the industrialised countries new rules for international trade favourable to them. Although the Uruguay Round was the longest, most ambitious and most complicated round of trade negotiations in the twentieth century, and even though eighty-one developing countries took part in it, during the last and decisive year all the delegations of those countries in Geneva found themselves having to wait for the United States, the European Union and Japan to reach agreement on all the areas under negotiation, without themselves having been able to intervene or to change anything in the texts that the rich countries had accepted. The negotiation became virtually an exercise among the TRIAD elite, aimed at harmonising the global interests of the transnational corporations. The text of the Final Act of the negotiations was presented to the developing countries as a *fait accompli*.

The countries formerly known as the Third World were always too politically, economically and culturally heterogeneous to join forces as a world front based on common interests. Now, their interests are still more diverse and distant, as the result of a greater differentiation in the levels of national misery among them. Divided into different groups with greater or less poverty, the majority of these countries have not begun to compete in the prosperity rankings with the old aristocracy of the industrialised countries. Instead they continue to compete among themselves for the world poverty ratings. This situation is a major obstacle for true economic globalisation, since it prevents the majority of the world's populations from becoming consumers of products from the transnational enterprises. For the devotees of Adam Smith's *The Wealth of Nations*, ironically, the concern today seems to be the poverty of nations.

Only a handful of the nations known as developing countries, such as Chile, Uruguay, Costa Rica, Malaysia and Thailand, have incomes per capita above $5,000 and enjoy the luxury of having less than 20 per cent of their population below the poverty line. A majority of the nearly seventy countries, like Brazil, Ecuador, Mexico, Peru, the Central American countries, and others from the Middle East and South Asia, on the other hand, have lower average incomes, and 25 to 40 per cent of their populations live in poverty. These, however, are not the poorest. At least thirty-five countries, including Nepal, India, Bangladesh, Haiti and the great majority of the African countries, have an income per capita of less than US$1,500 and have more than 50 per cent of their populations subsisting in poverty.[12]

Currently, the World Bank uses an international poverty measure, according to which the poverty of a country is determined by the percentage of the population that lives on less than one dollar a day. With this indicator, it is possible today to establish a world ranking of national misery. The country with the highest percentage of poor people is Guinea-Bissau, with 87 per cent of its population living on less than one dollar a day.

The twenty poorest countries in the world have between 80 and 30 per cent of their populations below the international poverty line of less than one dollar a day (see Table 2.1). They are followed by a sizeable number of developing countries that have between 25 and 29 per cent of their populations in this situation. Among these, it should be noted that there are even some countries classified as 'emergent', including China, with 29 per cent, Brazil, with 28 per cent, Philippines, with 27 per cent, and South Africa, with 23 per cent.[13]

TABLE 2.1: *Proportion of People Living on Less than a Dollar a Day (%)*

Guinea–Bissau	87.0
Zambia	84.6
Madagascar	72.3
Niger	61.5
Senegal	54.0
Guatemala	53.3
India	52.5
Kenya	52.5
Uganda	50.0
Peru	49.4
Honduras	46.5
Rwanda	45.7
Nicaragua	43.8
Zimbabwe	41.0
Botswana	35.0
Ethiopia	33.8
Mauritania	31.0
Ecuador	30.0
Nigeria	29.0
Bangladesh	29.0

In the year 2000, about 1.3 billion human beings were living in poverty. The greater part of the planet's poor in the twenty-first century will be concentrated in Asia (some 600 million), followed by Africa (450 million), and Latin America (220 million). If the situation does not change in the latter region, the countries with the most poor will be Haiti, Guatemala, Peru, Nicaragua, Honduras, Ecuador and Brazil.[14]

Since 1990, the UNDP has used a new formula, called the Human Development Index, which assesses national misery more accurately. This new formula combines GNP per capita with two other variables: life expectancy and access to education. It also takes into account profiles that measure the provision of basic human needs such as health care, housing, sanitation, drinking water, and food. This equation gives a truer view of real living standards than the mere growth of GNP, since it is often the case that GNP growth does nothing to improve the living conditions of the majority of the population.[15]

With this index, a world ranking of human development has been established for more than 174 countries, providing concrete evidence that poverty is the rule, and human well-being the exception. Listed in the highest ranks, with the highest degree of human development, are the whole aristocracy of industrialised powers, plus the Asian NICs (Hong Kong, South Korea, Singapore and Taiwan). The only so-called developing countries that are close in the rankings to the world industrial aristocracy are: Costa Rica, Argentina, Uruguay and Chile. Besides having higher per capita incomes than the rest of the so-called developing world, these countries appear at that level because they offer greater social integration, more education, and higher life expectancies. All the countries of Latin America, Asia and Africa that follow them in the ranking have very mediocre or low human development, with lower life expectancies and insufficient provision of the basic necessities like food, education and health care.[16]

In Latin America, the lowest human development indices are found in Bolivia, El Salvador, Honduras, Guatemala, Nicaragua, Paraguay and Peru. Peru is classified in the ranking below the majority of South American countries. It records poverty percentages above 40 per cent of the population, coupled with considerable infant mortality and malnutrition. The daily calorie intake per capita is the lowest of all Latin America, even falling below a good number of African countries.[17]

Three mega-countries that are considered as possible world

powers in the twenty-first century, Brazil, China and India, are badly placed in the human development ranking. In spite of its relative industrial development, Brazil has a low human development index in Latin American terms and a mediocre one in the world ranking. This is because large segments of its population are completely marginalised. Nearly 29 per cent of the Brazilian population, some 47 million people, live in poverty. The same is the case with China and India. Paradoxically, both are nuclear powers, even though they have a low level of human development and great sectors of their populations are living in poverty and neglect. China has 30 per cent of its population in poverty, some 359 million, and India has 52 per cent, or 520 million. These three giants of the so-called developing world will need to make enormous efforts to provide employment for their explosive urban populations and help them to free themselves from poverty to form a middle class majority. If these countries fail in this endeavour, the twenty-first century has serious socio-political disturbances in store for them.[18]

The World Bank's measure of poverty, based on an income of less than one dollar a day, and the UNDP's human development rankings confirm the existence of an enormous mass of national misery in the misnamed developing world. They also show that the huge numbers of people who are victims of low purchasing power and utter destitution prevent the formation, in the majority of these countries, of genuine market economies with national dimensions. No matter how much the economy is liberalised, deregulated and privatised, a great part of the population remains outside the national and the global markets. Poverty prevents the expansion of a large and productive middle class made up of consumers aware of their citizenship. That is a prerequisite for these countries to become modern, capitalist, democratic nation-states capable of integrating efficiently, along with the majority of their populations, in the global economy.

The billions of persons living on less than one dollar a day and with a deplorably low standard of human development are not the only, or even the major, obstacle that will prevent the majority of these countries from being integrated into the global economy in the future. Another impediment, as forbidding as their social misery, is their national scientific and technological destitution.

On the international scene, the scientific and technological development of a country is calculated by the number of scientists and engineers, the quantity of computers and expenditures in scientific–technological research and development. Today the misnamed developing countries, accounting for three quarters of the world's population (4.8 billion) possess only 10 per cent of the world's scientists and engineers. Of these, 7 per cent are in Asia, 1.8 per cent in Latin America, 0.9 per cent in the Arab countries and 0.3 per cent in the rest of Africa. These countries have only 3 per cent of the computers and invest only the paltry sum of 3 billion dollars for scientific research and development. In contrast, the world economic aristocracy, with one quarter of the world population (1 billion inhabitants) has 90 per cent of the world's scientists and engineers. Of these, 90 per cent are found in the United States, the European Union and Japan. This aristocracy also possesses 97 per cent of the computers and invests more than 220 billion dollars in research and development each year.[19]

The underdeveloped countries' national scientific and technological deficits will have serious repercussions on the viability of all these economies in the new century, since one of the new characteristics of globalisation is the changes taking place in the structure of the world demand for manufactured goods and services. The demand for manufactures and services with a high technological content is growing far more rapidly than the demand for raw materials and manufactures that are only slightly processed. The world demand for high-technology manufactures is growing at 15 per cent per year, while the demand for manufactures with low technological intensity is growing at only 5 per cent, and the demand for raw materials at between 2 and 3 per cent. Nowadays, enterprises are obliged continually to innovate technologically in order to remain competitive. The economies that cannot do this will be gradually excluded from the global market.[20]

The only countries that have companies that invent and export software are India, Hong Kong, Taiwan and Singapore. Companies from the latter three have branches in China, Indonesia, Thailand, Sri Lanka and the Philippines. These same NICs and, to a lesser degree, China, Brazil and India also participate in joint ventures with transnational corporations in high-technology industrial projects. Nearly all the remaining so-called developing countries are left on the sidelines, with negligible amounts of high-technology, or even medium-technology manufactures for export. Latin America is a barren source of exports with high technological content. The only exception is Brazil, with its modest computing, aeronautical and military industries.[21]

The technological modernisation that has become vital for any country hoping to compete in the globalised world will prove extremely difficult to achieve. This is the case not only because the study of applied science has been largely abandoned in the majority of the Latin American, Asian and African countries. It is also a result of the excessive protection of intellectual property exercised by the industrialised countries today, which makes it practically impossible for the poorer countries to copy and adopt foreign technology. During the Industrial Revolution, many of the now industrialised countries did not have to face international barriers such as these. Most countries at that time did not have national patent laws or, if they did, they were not connected to international commercial sanctions. This allowed countries to reproduce technologies. Nearly all machinery was copied, as were many chemical formulas, the automobile, the aeroplane, the radio, radar and thousands of other inventions, without a single country being called a pirate and threatened with economic sanctions.

Today, the growing scientific and technological deficit in the underdeveloped countries and the enormous difficulties in making it good reveal better than any other indicator that it is a euphemism to call these 'developing countries'. They are unable to compete globally to obtain resources from the world market so as to increase the income of a growing and expanding urban population that is trying to subsist on the feeble income from low-technology exports. The greatest challenge facing the misnamed developing countries in the twenty-first century will be to avoid the trap of scientific and technological backwardness resulting from their lack of historical cultural interest in scientific theory and applied science.

The huge expansion of poverty and the current indices of low human development, coupled with an enormous lag in scientific and technological knowledge, make the plight of the underdeveloped countries look very much like that of the third estate in the *ancien régime* of France. Indeed, these countries together constitute the majority of the world's population, just as the third estate was the majority within pre-revolutionary France. Just like the third estate in France, all of the underdeveloped countries are today at the mercy of the power of an aristocracy, the transnational aristocracy, and dominated by doctrine – of the supranational high clergy of the IMF and the World Bank. The great difference with the French third estate is that this international third estate has no revolutionary potential, does not constitute the embryo of a new middle class of newly industrialised countries with the power to negotiate with the new transnational aristocracy and with the supranational high clergy.

Due to their scientific and technological indigence, the great majority of these countries are caught in a real non-development situation. For decades they have tried to develop, using all sorts of models and ideologies, with no success in diminishing their poverty. After more than twelve years, they continue to follow with devotion or resignation the single creed of the IMF and the World Bank, granting all kinds of concessions to the transnational aristocracy with the hope of joining the global economy. However, the transnational investments and technology have not materialised, and their lack of technological modernisation prevents them from competing and increasing their income. Since they cannot work miracles with the scant income they obtain from exports (primary and with low technological content), to satisfy the growing needs of their exploding urban populations they have no option but to sink further into debt. The economic history of the so-called developing countries is nothing more than a series of growing debts and insolvencies.

The question that comes to mind is: when will all this national misery turn around? With the radical *laissez-faire* policies of the single creed imposed by the transnational aristocracy and the supranational high clergy, where everything depends on the invisible hand of the market, no one can possibly know. Perhaps it will occur when the critical mass of transnational investment and technology arrives or when the urban population explosion abates and, as is forecast, the developing world stabilises, in the year 2050, with some 8 billion inhabitants. In the meantime, it would not be surprising to see growing delinquency and socio-political disturbances in the international third estate.

CHAPTER 3

INTERNATIONAL DARWINISM

From Adam Smith to Charles Darwin

In *On the Origin of Species by Means of Natural Selection* (1859), Charles Darwin maintained that by a law of natural selection only the species fittest to compete for their survival can reproduce and indeed survive. Little did he suspect that some 140 years later, at the dawn of the twenty-first century, a global market and a technological revolution would apply the same rule, allowing only the fittest people, companies and national economies to survive. Thus those deemed less competitive are marginalised and seen as economically unfit species. The only difference between contemporary economic, technology-based Darwinism and the natural law is that the latter discarded the unfit species over millions of years, while the current market and technology-driven selection process can put thousands out of work within months, expel competing companies from the market in a couple of years, and take no more than a decade to make many nation states into non-viable economies.

By the mid-1980s, the Soviet system, based on a centrally planned economy in a pointless attempt to dispense with markets for a period of fifty years, began to collapse. At the same time, more radically *laissez-faire* policies started to prevail in the world, particularly driven by the more conservative forces in the Western world, typified by the Reagan and Thatcher administrations. When the Berlin Wall fell, in 1989, these policies took on fundamentalist features. The long-forgotten prophet Adam Smith was revived to remind all men on earth that society does not exist, that only the market is real, a market where individualistic self-centred interests oppose one another to create happiness for all.

Today, Adam Smith is viewed as the inspiration behind the new world economic order. In his best-known work, *The Wealth of Nations*, he tries to show that the pursuit of individual interest benefits society as a whole. However, this work is by no means a fundamentalist vindication of the market forces' prevalence in the economy. Contrary to the interpretation of radical advocates of neoclassical theory, Adam Smith believed in the importance of social aspects and moral concerns.

In *The Wealth of Nations* (1776), individual interests are portrayed as the driving force behind a successful economy, but what Smith refers to as individual interest is not selfishness or anti-social ambition. Smith, who was a moralist and a Scotsman, considered that personal interest was, indeed, the engine of a successful economy, but only if it was contained in a framework of social morals, which Smith called 'reasonable conduct'. Following the traditional Anglo-Saxon political thinking of Hobbes and Locke, Smith assigns an important role to the state, not only through an efficient legal and judicial system, but also as the entity responsible for guaranteeing workers' living standards. Smith thought that their living standards could deteriorate if the division of labour produced enormous material benefits but, through the individual's over-specialisation, impoverished their intellectual life. He feared that the latter could render them incapable of maintaining rational conversations or conceiving generous or noble feelings. As a result, he felt that workers would neglect their personal duties and the interests of their country. Therefore, for Smith, the state is duty-bound to resolve these human limitations, ensuring that each individual exercise 'intellectual and social virtues'.[1]

Prior to *The Wealth of Nations*, Smith published another major work, *The Theory of Moral Sentiments*. Its main theme is the proclivity of human nature to live in society and the need we all feel for our behaviour to be approved and appreciated by others. For

Smith, satisfying individual interests is encompassed by this ethical proclivity of man, so that the individual economic interests of a person or a company do not allow for the unbridled pursuit of profits regardless of the moral and social consequences of such pursuits. Smith was not an economic Machiavelli. He did not attempt to justify the predominance of individual interests at any cost, but rather within the boundaries of moral sentiments designed to control egotism.[2]

His successors the radically neoclassical economists and the ultra-liberal politicians overlook the moral and social climate in which, according to Smith, the economy should operate. The positive egotism to which they refer, and which disregards all social and ethical concerns, bears no relation whatsoever to the moralistic thesis argued by Smith in *The Wealth of Nations* and *The Theory of Moral Sentiments*. Partial quotations are taken from the former, while the latter is totally ignored.

Fashionable economists and politicians who revere Adam Smith have not only ignored the moral and social context in which these works were written. They have taken things a step further by trying to express in mathematical terms the metaphor of the invisible hand of the market, which Adam Smith used to show that, in a free market, balance is always restored between supply and demand, and which he believed guarantees that true consumer preferences are always satisfied. Modern neoclassical economic theory has formulated supply-and-demand mathematical calculations under the pretence of making economic policy into a pure science. Adam Smith never intended this metaphor – which appears only once in the 400 pages of *Wealth of Nations*, to become a hard-and-fast law of nature expressed in mathematical formulas. He never claimed to have proved that the market operates with the same degree of predictability as Newton's law of gravity.

To present economics as a pure science and the market as an ethically impartial law of nature, which decides which person, company or nation is fit to compete and which is not, regardless of the unemployment, poverty or degree of underdevelopment of each country, is to turn Adam Smith's description of economic freedom and moral responsibility into global market Darwinism. Indeed, the underlying neoclassical economic reasoning followed by the end-of-century ultra-liberal politicians instinctively uses the same basic axioms posited by Charles Darwin: dualism, conflict and evolution.

For Darwin, 'dualism' is the antithesis between the species, on the one hand, and the environment, on the other. Species are constantly adapting to the environment in an effort to survive. According to current neoliberal economic thinking, the world market is the natural environment to which one has to adapt in order to survive. Those persons, companies or national economies that fail to adapt are punished and pushed onto the sidelines as economically non-viable species. It follows that the market is not a human creation, but rather a natural environment, beyond our will: an invisible hand, devoid of moral judgement, a natural selectivity law, that can eliminate jobs, stores, companies, and make national economies non-viable. All problems will be solved by the natural market forces. They will select the persons, companies or national economies that are efficient, exactly as nature selects the most fit among the species, discarding the unfit.

Conflict, for Darwin, is the natural state in which all creatures live: as predators. This means survival and reproduction. Companies and national economies must also be predators (tigers), waging the fiercest economic competition. Only the most predatory economies prevail and reproduce transnationally, multiplying their growing returns. Financial speculation, although it generates no jobs, and technology, although it destroys them, must both become the means of achieving the highest possible profits and allowing for a mutation towards more economically fit and powerful species.

According to Darwin, evolution makes species change from primitive forms to more complex and subtle forms through various stages of mutation. This ability to mutate enables species to triumph and ultimately endure over the centuries. Well-adapted species can replicate themselves and reproduce. This scientific axiom, too, has been automatically transferred to contemporary economic thinking. Companies and national economies must adapt (innovate and develop) in order to conquer in the economic environment and prevail as efficient, viable economic entities. Companies can only evolve and reproduce through technological transformations so as to succeed on the global market, generating an ever-rising spiral of profits and prosperity. Countries that have not yet completed their evolutionary process will do so through the liberalised global market until they turn into modern, developed, capitalist economies.

The Darwinian concepts of dualism, conflict and evolution reflect cultural reflexes ingrained in the subconscious of westerners for generations. As such they resurface unfettered when social and economic events are considered. In this way, they serve to legitimate special interests and privileges. Concepts of dualism, conflict and evolution have been applied to modern economic and social reasoning both in Marxist rationalism and in the current neoclassical economic rationale that underpins capitalist globalisation. Both schools of thought refer to Darwinian equivalents because they both stem from the ideology of material progress spawned by the Industrial Revolution.

Marx uses equivalents of Darwinian dualism, conflict and evolution. That is how he justifies the class struggle, the ensuing revolution and thus predicts the victory of a new, fitter species: the proletariat. Indeed, he wanted to dedicate Volume One of the English translation of *Das Kapital* to Charles Darwin. His ideological disciples, such as Lenin and Stalin, were enamoured with material progress, and the idea of using the mechanisms of social accumulation and industrial production to foster the development of a new, fitter species: communism. The Soviet commissars became the omnipotent leaders of a megalomaniac drive towards industrial progress, which was responsible for massive human sacrifice, gulags and environmental crimes. The economic fundamentalism by which present-day capitalist globalisation is being justified is no less reliant on Darwinism. It is wedded to productivity and competitiveness, whatever the social cost, so long as the highest rate of consumption and individual accumulation of material goods is achieved. Production and productivity must increase, any pattern of consumption goes, environmental deterioration is not deducted from rates of return. The struggle and competition in the marketplace is natural, it is the course of life; the fittest person or company succeeds. The weaker go bankrupt or become unemployed. This is how, without the slightest pang of conscience, an analogy is established between Darwin's natural selection – which, like any natural law, is ethically impartial – and human economic activity – which cannot be morally neutral.

The underlying Darwinism of the neoclassical, ultra-liberal message that inspires current capitalist globalisation, turns the economy into the paramount factor determining all other options, whether political or social and even cultural; nothing could be closer to the Marxist ideology. However, the archetype is not the robot-like *homo sovieticus*, but rather the *homo economicus*, whose sole motivation is money, the ability to consume more material goods, who is aggressively competitive, a kind of predator loose in the Darwinian jungle of social and economic deregulation. In this jungle, not only companies but also individuals, each social group, each community, must be the fittest, the strongest, the best. Those who are not competitive must be eliminated from the economic arena, regardless of social, moral or environmental implications. This is a zero-sum game, where there is no co-operation. You win or you lose.

Dispossessed of their ethical dimension, the laws of economics, the market forces, are equal to laws of nature: amoral. Like the law of gravity by which the earth revolves, they operate regardless of whether we are good or bad, or whether lions prey on zebras. This whole ultra-liberal conception is closer to the biology of Charles Darwin than the economic policy of Adam Smith who, as already noted, ignored the moral significance of economic activity.

Smith would not have understood the rules by which the neo-liberal global game is played. These rules allow goods and capital to circulate and compete freely around the world, but set obstacles for people in search of a job. People cannot travel and compete freely: stringent immigration laws are freely enacted to prevent the globalisation of the labour market. In this way, neoliberalism does not apply to people – unlike Smith's liberalism, freedom does not extend to all the factors of production. The author of *The Wealth of Nations* would not be able to grasp, either, how economic growth can be attained without job creation. More impossible still for him to comprehend would be a global economy whose main driving force is the financial markets which produce no real wealth for the nations.

The global jungle

Nothing could be farther removed from the ethical and liberal principles underlying Smith's economic policy or closer to a jungle governed by Darwin's law of natural selection than the current process of globalisation. Indeed, the world market functions like a law of natural selection from which no one can escape, neither persons, nor companies nor nations. All must be resigned to adapt to this law, to accept unemployment, to witness entire sectors of the national economy being wiped out and to welcome the gener-alisation of consumer patterns that the planet's ecosystems cannot sustain.

In this global Darwinian jungle, the underdeveloped countries that fail to modernise technologically are not the only unfit species. The individuals and social groups that are less skilled in grasping and using new technologies are too. Thus 30 per cent of the global workforce are unemployed or underemployed, and poverty and social inequalities have worsened throughout the world. There are close to 100 million poor in the Western industrialised countries. If we add Russia and eastern Europe, the total of poor people reaches some 200 million. In the underdeveloped countries, there are almost 1.3 billion poor people. According to the United Nations, the ratio of the income of the world's richest 20 per cent to that of the poorest 20 per cent rose from 30:1 in 1960, to 61:1 in 1991 and almost 80:1 at the end of the century. The number of poor people has grown in practically every country, and the middle class's income has not risen, as it did during the sixties.[3]

The Achilles' heel of the world economy, as it turns into a global supermarket, lies in the underdeveloped world's lack of purchasing power. Of a total world population of about 6 billion, there are scarcely 1.8 billion consumers who can really afford products and services on the world market. The only products being purchased worldwide are those of the transnational entertainment industry, such as music, films, television series and cheap consumer products like jeans, trainers, cigarettes, fast food. Nowadays, only a minority of consumers in the world have access to new cars, computers, video cameras, digital phones, faxes, the internet, quality clothes, tourist trips abroad or international credit cards.

According to the calculations of certain transnational banks, of the 6 billion inhabitants of the planet, only some 900 million have enough income to be 'bankable' or to be offered an international credit card. Therefore the great majority are not 'bankable', have no access to international credit and thus cannot take part in the globalisation of consumption. Should this situation persist in the twenty-first century, these same bankers admit that it would jeopardise the future expansion of the global economy.[4]

Nevertheless, most transnational corporations consider that there is still enough untapped market potential in the world. What matters to them is that national markets become increasingly liberalised so that they can seek the thin strata with high income in the underdeveloped countries. India, once it has switched to the rules governing the game of global liberalised trade, with a population of over 1 billion will offer a market of at least 200 million people with adequate purchasing power, leaving the remaining 800 million Indians gazing at the shop windows. From the transnationals' standpoint, these 200 million people are equivalent to the entire US market (a mere trifle), enough to keep them in business for some time. This approach – of increased global prosperity with great social exclusion – is also applied to China. Transnationals do not aim to sell to the entire population. It would be sufficient for 300 million people in the upper income brackets, out of the total 1.2 billion Chinese, to become their customers, though this may create a dangerous gap between two Chinas, one marginalised and the other a consumer. Likewise, it is irrelevant that a major part of the population of Brazil, or Indonesia, or Russia, or Mexico or the Philippines is marginalised; what matters is capturing the high-income minorities.

Another feature of the global jungle is not only that it produces more with fewer jobs for a minority of the world population, but also that it has created a vast planetary gambling house. Every day, in this world casino, a number of creative games in financial speculation are played, involving investors, mutual funds, stock exchanges, currency markets, bonds and securities. Billions of dollars' worth of transactions are made on a daily basis, trillions per year, mainly bearing absolutely no relation to trade or to direct transnational investments to create jobs. A week's worth of speculative transactions on the world currency markets is virtually equal to the aggregate total value of international trade and foreign investments in a year.

This gambling house never closes. When trading stops in Europe, it opens in New York, followed by Tokyo and Hong Kong. By means of a worldwide satellite-based telecommunications system, thousands of brokers, bankers, investors, capital managers, executives and even private individuals, instead of playing poker, blackjack or roulette, through their computer screens bet billions of dollars on a global scale, buying shares, selling bonds, undermining the value of a currency, fixing a futures price, playing with derivatives and performing a variety of sophisticated and volatile financial speculations. The point is to make a lot of money, and fast.

The main difference between this world casino and an ordinary casino is that it impacts even on those who do not bet. A change in a currency value or speculation with the price of a raw material can produce huge gains, but can also cause the flight of capital or a crash on the stock exchange, and can drive companies to bankruptcy and lead to dismissals.

The world casino has had devastating Darwinian effects. In the mid-seventies, it recycled petro-dollars by offering soft loans to the so-called developing countries; in the eighties, it created the conditions for a rise in interest rates, thus unleashing a debt crisis that made the Latin American countries non-creditworthy. After 1990, investors in the world casino became more predatory. Nowadays they move in packs, invest heavily, make big profits, are extremely ambitious, and fly into panic at the slightest hint, not of losses, but of diminished returns. They tend to invest disproportionately in countries that become their favourites but, at the first cloud, they flee in a stampede leaving their previously cherished emerging currencies and economies devalued and devastated. In 1990, they attacked the European Monetary System, forcing the pound sterling to devalue. In 1994, a capital stampede made Mexico insolvent once again; in 1997 a capital stampede pushed into bankruptcy no less than the emerging economies of Asia.

The mega-financial crisis that almost turned tigers such as South Korea and Malaysia into vegetarians and that inflicted harsh punishment on such (undeservedly called tiger) underdeveloped economies as Thailand, Indonesia and the Philippines stemmed from an enormous tsunami wave of speculative investments headed by Japanese, US and European investors. Foreign investment, which for decades had been channelled towards creating modern factories and improving the technological content of exports, turned into speculative operations in 1990. On account of the deregulation of financial markets, the lack of democratic tax regimes, and corruption, capital was now used for speculative loans invested in real estate and in the creation of innumerable banks and financial institutions; such loans even went to pay back personal favours by financing non-profitable businesses. When the borrowers started finding it difficult to pay back the casino investors, there was a stampede. Dozens of billions of dollars from the central banks of Thailand, Malaysia, South Korea, the Philippines and Indonesia were engulfed by the speculators and transferred to private financial institutions. All these countries were compelled to accept the supervision of the IMF and to apply austerity measures that hit the personal incomes of millions of citizens who had never indulged in playing Asian roulette.

In Latin America, casino-style investment has been much lower than in Asia. Moreover, it did not even go through the phase of financing modern factories and technological innovation, as it had in Asia. No, in Latin America, from the outset, investment was Darwinian and highly speculative. Here, the bulk of global capital flows have concentrated on securities, stock and shares and, to a lesser extent, on acquiring assets through monopolistic privatisations, leading to rapid recovery of investments. The scarce productive investments focused on raw material exports of low-

technology products. Thus, the casino investments have compounded the Latin American countries' position as exporters of
raw materials or low-technology manufactured goods. As a result,
current account deficits in Latin America are accumulating and,
adding insult to injury, rely on short-term speculative capital from
the global casino to be financed.

Latin America's chances of winning in the world casino are
slim. It would need to be dealt a handful of aces, which has not
occurred up to now, so that short-term speculative capital could
form a critical mass of productive foreign investment that might
give rise to modern enterprises, create jobs and increase the technological content of exports. In the wake of events in Asia, this is
highly unlikely. What is more likely is that, at the slightest sudden
increase in current account deficit increase, or sign of devaluation,
or political instability, or populist measure in Latin America,
foreign capital will dry up or take flight and, once again, render
many countries in the region insolvent, unable even to finance
their present growth, which is happening without job creation.

Capital in this global jungle is very jittery. It is apt to flee from a
given country in a flash. Besides, it represents an almost virtual
value, which no government in the world can control. The
world's economic power has shifted into the hands of the big
financial predators, whose appetites and stampedes cannot be
tamed by even the most powerful central banks. The heads of the
Group of Seven richest countries in the world can only publish
communiqués expressing their concern, and the IMF and the
World Bank have no authority to regulate the operations of the
global casino. These two institutions instead concentrate on giving
poor countries recipes to attract the increasingly volatile foreign
capital which is not a source of jobs and which escapes even their
control.

Not only is the global jungle a financial gambling house, it also
generalises consumer patterns that are incompatible with the

planet's ecological balance. All contemporary consumer patterns are based on an economic model in which nature is viewed merely as another raw material to be consumed. Farmland is taken over by urbanisation, reducing food production. Over-fishing is depleting fish stocks. Greenhouse gas emissions create air pollution leading to climatic changes, droughts and floods. Chemical and nuclear industries continue to bury their toxic waste. Goods are shipped across the world wrapped in kilometres of paper and cardboard derived from deforestation.

With the globalisation of consumer patterns, rubbish is also globalised. Mountains of nuclear, chemical and toxic waste, along with syringes, bandages, hospital gauze, paint, plastics, refrigerators, old tyres, clothes, metal objects, ceramics – all the things the industrialised countries discard – are systematically shipped to new global rubbish dumps in the underdeveloped world and in eastern Europe. What would happen if the same consumer-based prosperity were achieved in the underdeveloped countries? If globalisation were to succeed, would it be possible to recycle the additional rubbish produced by 4 billion new consumers? Or would we have to start changing our consumer patterns?

A baby born in the United States is double the burden on the environment that a Swedish baby is; 3 times that of a child born in Italy; 13 times that of one born in Brazil; 35 times that of an Indian baby; 140 times one born in Bangladesh and 280 times what a child costs the environment in Chad, Rwanda, Haiti or Nepal.[5] Can the environmental cost of the American birth, which is the model conveyed by modern consumer patterns, be extended to the 80 million babies born worldwide each year?

If industrialised societies' consumer patterns are globalised, the earth's biosphere will be unable to sustain them. It is an established fact that widespread damage to the environment started with the Industrial Revolution and that its pace has accelerated exponentially with the emergence of contemporary mass consumer societies. The aim of the world jungle, however, is nothing other than to extend the unsustainable consumer patterns currently followed by some 1 billion people, to the almost 5 billion inhabitants of the 'developing' world. The lifestyle in the highly consumerist societies is promoted throughout the world. The irony lies in the fact that unemployment, poverty and marginalisation prevent its spreading, otherwise the earth's biosphere would be rapidly destroyed by 5 billion international credit cards.

According to the present transnational globalisation model, the planet is far from becoming a world village in which all nations partake in prosperity and nature conservation. It is rather like a string of planetary ghettos in which rich people, the consumers of the global economy, share the same lifestyle and destroy the environment. The inhabitants of Beverly Hills and the rich neighbourhoods of Mexico City, Lima, Johannesburg or Bombay, despite the continental distances between them, live in very similar conditions, in contrast with the most dissimilar conditions of the surrounding neighbourhoods. From Los Angeles to Vladivostok and from Rio to Manila, the poor, unemployed majorities, with no prospect of increasing their income, live next door to small elites in their walled-in properties, patrolled by private police forces, consuming all sorts of exquisite global goods.

This Darwinian global village has a high street made up of the elegant ghettos of all the cities in the world. Slums, shantytowns and hovels proliferate in the back streets. In these new global urban areas, the largest and most populated human settlements, water, energy and food are scarce; sweatshops mushroom, along with air pollution, illegal work, child exploitation, unemployment, prostitution, petty crime and delinquency. Very soon from these precarious and unhealthy global settlements, teeming with human energy, informal sector activity and unemployment, will spring the fate of many poor countries, for they will produce the new politicians, businessmen and professionals, and they will spawn the delinquents and extremists who will threaten the established order.

Adjustment without modernisation

The present situation is the closing instalment of the Darwinian saga that, as noted above, started in the mid-seventies when a huge supply of credit was caused by the enormous surge in oil profits deposited in transnational banks. The bankers wasted no time in travelling throughout the Third World, offering loans. The governments of the underdeveloped countries, struggling to pay for the increased costs of all their imports due to the soaring oil prices, had to choose between the temptation of the bankers' offers and plunging their economies into a recession and becoming unpopular. They obviously opted for the former.

Citicorp was one of the main banks that promoted this loan frenzy. Its President, Walter Wriston, one of the boldest bankers on the stage today, invented the 'recycling of petro-dollars' accompanied by the motto: 'States never go bankrupt, they always pay.' He was right. The debt crisis destroyed a decade of social progress, but the states paid up. The banks collected their interest through refinancing schemes based on stringent, IMF-supervised adjustments. At the start of the 1990s, the situation was under control. The debtor countries, under strict adjustment and socially devastated, were paying US$50 billion per year.[6]

Once the danger to the banks had been eliminated, the next step was to throw the underdeveloped debtor economies into global competition. This was accomplished by means of the World Bank's structural adjustment programmes (SAPs), which were applied as the only way for the underdeveloped countries to establish their creditworthiness, and to avoid becoming international outcasts. This structural adjustment consisted in a kind of 'big bang' liberalisation, deregulation and privatisation, exposing in short order the underdeveloped countries to global economic competition, aiming to make them efficient capitalist economies.

By connecting primary, technologically backward national economies with the global economy by means of a rapid and indiscriminate liberalisation process, the IMF and World Bank made these countries produce according to their established comparative advantages in the international market. Thus the underdeveloped countries again produced what already existed, a traditional, primary export sector. There was no stimulation of modernisation, or diversification towards new, more technologically intense manufactured products and services.

In fostering structural adjustment, the World Bank was firmly convinced that, once economies were liberalised, the developing countries' comparative advantages in terms of cheap labour would attract the foreign investment needed to modernise them. The facts have shown that this belief was merely a theoretical notion, with no basis in reality. In the first place, given the speculative nature of the global economy, obtaining productive foreign investment is very difficult. Today's international financial market is a huge casino, processing transactions worth one trillion dollars on a daily basis. This is equal to nearly six times the value of all the direct productive foreign investment during one year in the whole world.[7] These transactions produce greater short-term gains than any type of industrial investment. In the second place, the transnational manufacturing corporations do not invest solely on the basis of cheap labour; they also seek good infrastructure, a broad internal market, a skilled workforce and local enterprises with sufficient technological capacity to produce parts and components for them, according to the rigid specifications of the global market. For that reason, more than 70 per cent of productive transnational investments are concentrated in the industrialised countries themselves.

With neither transnational investments nor technology, it will be virtually impossible to modernise the backward, primary-exporting economies. In spite of this, the IMF and the World Bank have made no systematic effort to promote contacts between the transnational corporations and the businessmen of the countries that have made structural adjustments, with a view to exploring the possibilities of productive foreign investment to modernise their exports. There is no international meeting place where the businessmen and the authorities of the countries that receive a minimum of foreign productive investment can negotiate with the transnational executives. The severe readjustments imposed by the IMF and the World Bank have pitted these unfortunate countries against neoliberal theories rather than against the real actors in the global economy.

From this Darwinian experience, it has become increasingly evident that merely liberalising, deregulating and privatising does not guarantee the formation of a critical mass of investment capable of modernising technologically the primary, backward economies. No one who knows the history of the modern industrialisation of the United States, Europe, Japan or of the Asian NICs can believe that technological modernisation can be achieved without promotion by the state. The United States itself presents the most persuasive example of promoting 'industrial policy' in the area of defence procurement, by which for more than half a century it secretly subsidised its industrial growth and ultramodern technological development.[8]

In the wake of the more than three hundred adjustment programmes that have successfully stabilised the underdeveloped economies while wreaking enormous social devastation and failing to modernise them, the World Bank has apparently executed an about-face and accepted the role of the state. The World Bank now not only dictates an economic adjustment policy that does not modernise the unfortunate underdeveloped countries, but pretends to emulate Hobbes, Locke, Bodin, Montesquieu, Lenin or Aron in contributing to the theory of the state. It indicates how the ideal state should work so that its structural adjustment programmes are not paralysed by bad governance or, above all, by corruption.

The World Bank seeks to rebuild 'state governance'. It advocates not dismantling the state, but modernising it to protect the most vulnerable social groups and to invest in social services, infrastructure and environmental protection. The World Bank advocates the modernised, democratic state, respectful of independent judicial power and free of corruption. But the World Bank's proposals are trite and contribute nothing new. Any economy will work better if the government is efficient and not corrupt. However, this needs to be taken further, granting, for instance, a modern managerial role to the state so that it may support national capitalism and help it to modernise and make its enterprises competitive, as in the cases of Japan, South Korea or Taiwan. The World Bank says nothing about this managerial role for the state. It would appear that it continues to leave capitalist modernisation to the market's invisible hand.

After ten years of neoliberal reforms, Latin America has not built up competitive momentum. Its national savings are insufficient, its exports are not diversified, it lacks capability for industrial innovation, its system of scientific and technological education is deplorable, its technological development is incipient, its infrastructure is inadequate, and its poverty does not diminish. Only the mirage of economic development persists, due to some short-term privatisations and investments. These enable such countries to import luxury goods for the high-income sector of the urban population, causing the majority of the people to dream of a prosperity attainable only for the very few.

In Africa, the results of adjustment policies have been more deleterious and even less certain. Africa is the developing region in which the World Bank's structural adjustment programmes have been most generally applied. Since 1982, when the debt crisis began, in Africa alone 162 programmes have been implemented, compared with 126 in all the rest of the developing world. Some African countries applied as many as four adjustment programmes between 1983 and 1993.[9] In contrast to Latin America, Africa has not attracted even short-term speculative foreign capital. On the contrary, its status as exporter of practically nothing but raw materials and basic commodities has been reinforced under the structural adjustment programmes. The region, as a result, exports a greater volume of such products, at lower prices, than previously.[10] Adjustment has, then, enmeshed Africa even more tightly in the basic products trap, with the added complication that Africa has the highest population growth rate in the world. After twenty years of undergoing an ideological experiment, Africa has entered the new millennium as a continent full of dysfunctional national economies out of sync with the global economy.

Despite their technocratic jargon, the World Bank's structural adjustment programmes have proved to be experiments heavily dependent on ideologies based on the Washington Consensus (see page 56) and on theoretical elucidations of the neoclassical economics that attempt to convert underdeveloped economies into modern capitalist ones in the image of the Anglo-Saxon model of capitalism. The IMF and World Bank officers are ideologically obsessed with the total deregulation of the economy, whereas the true problem of modernisation in the majority of these underdeveloped countries is historical and cultural. They lack a capitalist ethos and a scientific tradition needed to modernise their production through technology.

Today, for the most part, the underdeveloped countries are not modernising. They continue without technological change, without significant diversification of their exports, without creating companies with the comparative advantages required for mainstreaming in the hyper-competitive global market forecast for the twenty-first century. The majority of these countries have liberalised the economy, while continuing to export what they have always exported – raw materials and products with little technological content. Opening their economies has only increased their imports of manufactures, services and foodstuffs from the industrialised countries and the transnational corporations, thus creating new deficits that, some day soon, they will not be able to cover either by privatisations or by the short-term foreign capital accessible to them. Adjustment therefore represents the failure of Darwinian economic logic.

Deproletarianisation

Guided by a Darwinian market, technology acts as the decisive factor in discarding those who do not adapt to its progress. Today's technological revolution is eliminating the huge factories and their chimneys and the proletarian populations that ran them, and replacing them with smaller production centres that are substantially automated, rife with computerised information and temporary jobs. This type of modernisation is occurring in nearly all branches of industrial production. The result is an unstoppable process of world deproletarianisation.

This shift from a world economy based on factories to one based on intensely computerised enterprises is more radical and traumatic than the nineteenth century change from agriculture to factories. The labour-intensive Industrial Revolution was more gradual, allowing time for the agricultural society to adjust to the machine age.

In addition, the Industrial Revolution created more jobs than it destroyed. In contrast, the computer revolution can destroy more jobs, and do it more violently. It will not necessarily create better jobs for the large majority of workers, and it will deepen social inequalities. It happens more rapidly, and its social effects are more drastic. The unskilled workers are simply discarded. If those who lose their jobs and those who are not trained in the new technology do manage to find employment, they have to be satisfied, in many cases, with lower salaries than before.

This situation can be observed both in the United States and in Europe. Despite the fact that many jobs have been created within the technological revolution in North America, most of them are low-paying. Many of the new employees are called the 'working poor', that is to say, they become poor by working. The Nobel prize-winning economist Robert Solow, of the Massachusetts Institute of Technology, considers that although the United States has recovered virtually the same employment level as 1973 (the golden age of American employment), it has not recovered its level of well-being, because American society has become far more unequal in its income distribution, due to the enormous salary gap. On the contrary, not many new jobs have been created in Europe, partially because Europeans are not resigned to reducing their income, and prefer to wait and live on the unemployment insurance provided by their social security system, which is more generous than the North American version.

New technologies can erect new and rigid social barriers, especially between the skilled and the great majority of unskilled workers, or those who are only trained to perform precise computing jobs or temporary services that do not produce the level of personal income enjoyed previously. Felix Rohatyn, the banker who rescued the city of New York from bankruptcy, confirms this situation in his description of his country as a 'ruthless economy'. He also holds that in the US society an enormous transfer of wealth has occurred from the under-skilled working middle class to the capital owners and a new technological aristocracy.[11] The violent social upheaval produced by the technological revolution is not the exclusive province of the United States. It is occurring in all the industrialised countries. In some cases, such as Europe, it contributes to unemployment; in others, like the United States, it creates inequality.

Modern production is displacing the proletariat. There are fewer jobs per unit of production. Within the labour force, the proportion of common labourers has diminished in recent decades. According to a study by Professor Peter Drucker, in the United States and in Europe in 1960 there was one labourer in every four employed persons. Now the proportion is one in seven. The technological revolution has not stopped at the factory door, but has also invaded the office. By extension, deproletarianisation has penetrated the big bureaucracies, eliminating them by dint of software.

As a result of deproletarianisation, there is now a marked reduction in the unionisation of workers and employees. In the United States alone, the size of the unionised workforce has diminished by more than one third. It is also diminishing in Austria, France, Germany, Italy, Switzerland and Britain. Both in the industrialised countries and in the poor nations, the unions have lost strength and political clout. Today only 17 per cent of the world labour force is unionised. The first Industrial Revolution, with its labour intensity, made a political force of the proletariat. With the present technological revolution, the only remaining political forces are capital and software.

One of the most effective ways of deproletarianising production in the industrialised economies is 'lean production', consisting in splitting up a large factory into small workshops, using technology intensively, and very few rapidly trained workers on short-term contracts and with lower salaries than those of the previous (specialised) workers. These workshops receive orders by computer from the main factory, manufacture the parts rapidly, and send them back just in time to be assembled, thus saving the high costs of maintenance of stock at the central plant.

Another very efficient method is called 're-engineering'. It consists in the intensive use of computers to eliminate a whole chain of activities within the firm. A company that applies re-engineering installs a technician in the plant, who uses software performing an entire chain of human activities that were formerly carried out within the company. This new technician, called the 'dead structure', replaces with his software a complete chain of personnel, from professionals to secretaries. This method has increased productivity and the company's profits from services, and it is one of the methods that destroys the highest number of jobs.[12]

Lean production and re-engineering have massively eliminated jobs, even though they have managed to reduce costs and increase the profits of many major enterprises, and above all, the salaries of their top executives. The increased value of many companies' stocks and of the highest executives' salaries have been achieved at the expense of pyramids of professionals, technicians, secretaries and labourers who have been dismissed. Formerly the pride of a big executive was to create jobs; now the best-paid executives are those who enforce lean production and apply re-engineering.[13]

Together with these deproletarianisation methods, the fashion in world production today is to have a 'flexible' labour force. This means contracting personnel for one year or for six months, with a few hours of work per day, depending on the production requirements. The hours may be increased, but the worker may also stay out of work for weeks without earning any salary, waiting to be called to work. Under the terms of these new industrial contracts, there are neither rights to unionise nor rights to social compensation for dismissal.

If, in the United States and Europe, where population growth has been stabilised, the technological revolution has eliminated millions of jobs, or has created for the most part temporary, lower-paying jobs, what are the prospects for future employment in the underdeveloped countries? They are already suffering far greater unemployment rates even than that of the United States, Europe and Japan during the Great Depression of the 1930s. In addition, the growth rate of their urban working-age population is twice that of the industrialised countries at this moment.

The underdeveloped countries' population grows at a rate of nearly 70 million a year and the number of new job-seeking youths each year reaches nearly 38 million. That is, a population equal to that of all the Andean countries combined is born every year, and a population the size of that of Colombia is looking for work. Today the underdeveloped world has 700 million unemployed or sub-employed workers. What will happen in eighteen years to those who have just been born? What future can they have? The United Nations and the ILO consider that, if the urban population continues to grow at the same pace in those countries, one billion new jobs will be needed over the first decade of the new century.[14] How is this mass of world population to be employed, under the new technologies?

It will be very difficult to provide jobs, because nearly all the major transnational investments that use large numbers of workers (of the assembly plant type) were already made in the period 1970–1990. Moreover, it is very possible that, as the twenty-first century gets underway, new technologies will be gradually introduced in these same plants in order to cut costs, with the result that personnel will be further reduced or eliminated. The reason behind this move would be that it would be less costly and politically less risky to produce the same products in automated plants in the industrialised countries. In any case, if the employers wish to modernise their enterprises, improve their productivity and the quality of their products in order to become competitive global exporters, they will have no choice but to invest in new plants and more technologically advanced machinery, which will not create sufficient jobs.

New technologies and modern methods of production may possibly, after a painful transition period, reduce unemployment and low-paid employment in the wealthy countries, whose populations are practically static and generally skilled. However, they can hardly be expected to provide employment for the millions of unskilled men and women launched on to the labour market every year by the urban population explosion in the poor countries. Everything seems to indicate that the technological revolution and the urban population explosion in the underdeveloped countries will enter on a collision course in the new millennium.

Today many underdeveloped countries have reached the stage that the specialists call demographic transition, that is, a decline in fertility and a better balance between births and deaths. Nevertheless, there is still an explosive growth of the urban population. The cities of the underdeveloped countries grow at a rate of 150,000 inhabitants per day. Lima and other Latin American capitals grow by more than 2.5 per cent yearly, and other cities in Asia and Africa reach a rate of 3–4 per cent. This growth will be enormously difficult to fit in with the employment opportunities offered by the new technologies.[15]

This means that millions of men and women in the urban areas of underdeveloped countries are entering the global employment market in fierce competition for an ever-smaller number of jobs. Many will lose their jobs, or will find a low-quality job, or will never find work at all. In the underdeveloped world today, being a labourer, an employee or a professional is practically a privilege, and will continue to be so for many long years to come. In the poor mega-cities around the world, thousands of young people are wondering what to do. It is, therefore, not surprising that the answer should be emigration, delinquency and also political extremism.

The present structural trend away from the creation of sufficient jobs to meet the demands of the urban population explosion in the underdeveloped countries is operating within a planetary casino, hyper-speculative economy, disconnected from industry and trade and, as a result, from the creation of new jobs. In the course of this century, the combination of both these trends may consolidate a phenomenon made up of economic growth and insufficient job-creation. The problem of the world financial casino might be controlled through international co-operation, at least to establish regulations to prevent stampedes of capital and national bankruptcies like those in Asia. Technological development, however, is irreversible, as there is no such thing as disinvention, and the tendency of the present-day technological revolution is to dispense with the abundant labour force emerging from the urban population explosion in the underdeveloped countries.

An interesting example of this trend is that the 40,000 transnational enterprises and their 250,000 subsidiaries, which are the ones that use state-of-the-art technology, employ only 22 million persons in the entire world and of these only about 7 million are workers in the poor countries. They account for less than 1 per cent of the workforce of those countries. The transnational corporations are developing labour-saving technologies that are well adapted to the industrialised countries where they concentrate their production, but not to countries that are undergoing an urban demographic explosion. Although robotisation has not yet significantly affected unemployment, robot sales are rising, from Singapore to the United States, at 20 per cent per year. For each 10,000 workers in the United States there are 30 robots, in Sweden 73, in Japan 324. In future, this trend could gradually become a serious problem for the developing countries, exporting as they do labour-intensive manufactures.

Dematerialisation

Another Darwinian tendency of the market and of the technological revolution is the gradual elimination of national economies by the dematerialising of modern industrial production. The demand for raw materials is declining and their prices are always unstable and barely profitable. This is true because the new technologies use less and less raw material and fuel for each unit of production.

Today new technologies such as computerisation determine the exact quantity of metal or fuel needed; moreover, new artificial materials are replacing metals and natural textile fibres. At the same time, biotechnology is creating agricultural goods to compete with natural products.

The wealth of nations has changed in recent decades. Any classical text on international economics would have considered rich those countries with abundant natural resources. Today, the technological revolution is progressively emancipating the industrial economy from natural resources. The modern-day wealth of nations is information, the grey matter to create and innovate products and services, so as to save on natural resources. The key for this is software, which has become the new, strategic raw material.

The most important dematerialisation in progress is that of minerals and metals, which are being replaced by artificial materials. Among these new materials, designed by software, are plastics covered with very little metal, thermo-elastic plastics, and new forms of laminated glass, ceramics, polymers and graphite. All of them are in increasing use in the modern aerospace, electronic, chemical, telecommunication, computing, automobile and machine industries. Nowadays metal is replaced by ceramics and plastics in modern automobiles, aeroplanes and trains in order to reduce their weight, and thus use less fuel.[16] A large part of the energy that used to be lost in copper conduction cables is saved today through the use of new materials in semiconductors such as coaxial cable and optical fibres. Nowadays, 40 kilos of optical fibres transmit as many telephone messages as a ton of copper used to do. In the laboratories of Los Alamos they are experimenting with a new superconducting tape that can carry 1,200 times as much energy as a copper cable.

The amount of metal and minerals per unit of industrial production today is two fifths what it was in 1900. The raw materials used to make a semi-conductor microchip represent only 1 to 3 per cent of the total production cost. In the automobile, this proportion has been reduced by 40 per cent, in household appliances and in medicines by 50 per cent. The experience of Japan, which used to consume raw materials at a great rate, is illustrative. For each unit of industrial production today, Japan uses 40 per cent less raw materials than in 1973.[17] In the twenty-first century, this dematerialising trend in metals will have a growing impact on the countries that produce copper, aluminium, steel, tin, zinc, lead and iron.

The amount of energy needed for each unit of industrial production is also being reduced. Today, experiments are being conducted on a new type of energy, called a 'fuel cell', which obtains energy by combining oxygen with hydrogen in a kind of battery, where a slow reaction is generated producing constant electricity and heat. This new, clean energy is destined to replace gasoline for the combustion engine. The world's three largest auto manufacturers – Mercedes Benz, Mazda and Nissan – are experimenting with this fuel cell. If the production cost can be lowered by a few dollars, automobiles with combustion engines will be replaced gradually over the next fifteen years.[18]

Textiles are becoming the creatures of chemical technologies. Synthetic fibres are reducing the amount of cotton and wool in each unit of industrial textile production. Today there are already artificial microfibres so light that a kilometre of their thread weighs only one gram. From them, clothing can be made that is extremely light, like a second skin, and that adapts to the atmosphere, maintaining the temperature, heating or cooling, and preventing perspiration. During the Gulf War, these new fibres were tested in uniforms that allowed US soldiers to endure temperatures of 50 degrees centigrade.

The new chemical technologies, along with biotechnology, are also replacing agricultural products such as sugar, rubber, vanilla, palm oil. The market for sweeteners and artificial fats is being developed rapidly by the laboratories of the United States, Europe and Japan, thus reducing the demand for sugar and palm oil. Experiments are also under way to produce a bio-coffee, capable of competing with the best-quality coffee in the world, and at lower prices. The new, artificial vanilla is threatening thousands of growers in Africa. In the same way, research is being carried out to alter the DNA of flowers and tomatoes so as to make them cold-resistant and thus more suited to growth in the northern hemisphere. Should this succeed, it would reduce the demand for these products from the underdeveloped world.

The dematerialisation of industrial production in the developed countries is causing a decline in the demand for raw materials and lowering their prices to unprofitable levels. World Bank studies indicate that the real prices of commodities have plummeted, to a level even below the prices in 1932, during the Great Depression. According to the World Bank, the prices of all the basic mineral and agricultural products will continue to decline for the next ten years and, perhaps, even further into the century. While this is going on, the populations of the majority of the underdeveloped countries that export these products will have nearly doubled.[19]

The endowment of certain nation-states of the Third World with abundant natural resources, and the growing demand for raw materials accustomed their governments to live on that income, with no thought for scientific research. Nowadays, as the world economy is engulfed by the technological revolution, the economies of countries sitting on gold mines of natural resources work less and less well. Even the countries that are great world exporters of petroleum – the only raw material that is strategically vital today – such as Mexico, Venezuela, Nigeria, Iran or Saudi Arabia, have suffered severe crises and are implementing national austerity policies.

Today, petroleum is the only basic product that still provides a relatively high income. But the petroleum-producing countries, though they earned abundant capital, missed their historic opportunity to use it to modernise their economies technologically. The present industrialised countries and the Asian NICs, albeit not endowed with natural resources, have attained higher living standards than those countries sitting on 'gold mines' of petroleum, copper, bauxite, sugar or coffee. The only industrialised country that has enjoyed abundant natural resources has been the United States, although it owes its success as a world economic power not to them but to the grey matter it has used in order to invent and to innovate. Switzerland, a country about the size of a small department of Peru, has practically no natural resources, but it has attained great scientific and technological development. This allows it to sell industrial plants, chemical products, optical goods, watches, precision instruments, food products, sophisticated financial services, which produce infinitely more than all Peru's natural resources. In the last resort, Switzerland sells grey matter – an intangible, non-material resource contained in the technological innovation of its products and services. For that reason, it is a far more prosperous country than any country of Latin America, Asia or Africa that sells mountains of minerals or of woods, tons of fruits, sugar, coffee or kilometres of natural fibres.

The grey matter incorporated into software will be the most important factor in technological mutation to produce sophisticated goods and services and establish enterprises with comparative advantages for the twenty-first century. In order for a country to be viable, it has become increasingly vital for it to have an economy based on enterprises that are capable of building up the technological content of their production. If not, if they maintain the pattern of primary exports, the new technology that dematerialises production will perform its Darwinian selection function.

The Darwinian process of dematerialisation through technology will intensify during this century. The demand for manufactures with high technological content and, above all, for services that use little raw material, will grow at more than 15 per cent a year, while the average world demand for primary products will grow by only 2 per cent, and for slightly processed products, by just 5 per cent a year. The only commodities that will maintain an acceptable demand in the twenty-first century will be cereals and petroleum, in response to the growing demand for food and energy created by the world urban population explosion The world is entering the twenty-first century with a 'dual' planetary society, divided between a prosperous minority of persons and of countries dedicated to dematerialised intellectual activities, inventing modern technologies and new products and services, on the one hand, and, on the other, a majority of poor persons and countries that still live by dint of their physical strength, by routine bureaucratic work, and by tapping their natural resources.

CHAPTER 4

THE SEARCH FOR EL DORADO

Thinking the unthinkable

Theorists who cogitate about the wealth of nations and technocrats who specialise in formulating projects to increase production and raise living standards may be mistaken about the design of a development model, but they do not appear to entertain the slightest doubt as to the chances of development itself. For them, to consider the impossibility of development is to think the unthinkable.

Certainty about development has even led them to change the denomination of the 'poor countries'. Before the appearance of development theories, the poor countries that had not experienced the capitalist industrial revolution were called 'backward countries'. Then, in the 1950s, when development theories began to be discussed, the term was changed to 'underdeveloped countries'. Some time later, in the 1960s, the name was again altered, to 'countries undergoing development', because this indicated that they were on their way to a high standard of living. However, this seemed to imply some doubt as to their achieving the goal at the end of the road. Thus, the term was again adjusted, to 'developing countries'. This new denomination aimed to allay any doubt, by indicating that these countries were indeed on the road to material progress and high living standards. They seemed to be species that were being genetically reproduced on the pattern of the industrialised countries, assumed to be their ancestors in the history of material progress. In this way, development was represented as a natural process, like a Darwinian evolutionary certainty – the backward countries were 'developing', using the genetic potential of any nation-state to turn itself into a society with high living standards. The myth of development was born.

Development was one of the most persistent myths of the second half of the twentieth century. Theoreticians, experts and politicians have been convinced that economic and social development is an inborn, one could say inevitable, process for all nation-states. They think that it is only necessary to apply the correct theories and policies and poor countries will begin to create wealth and become societies with high living standards, like those enjoyed today by the twenty-four capitalist, industrialised democracies. Over a period of half a century more than 130 countries have attempted to apply various different economic and social ideologies and systems in search of development, as though that were El Dorado, the land of gold. Development, however, has proved to be as elusive as the conquistadors' dream.

The myth of development's origins lie in our Western civilization's ideology of progress. This ideology in turn was born during the Age of Enlightenment, and was fostered by the Industrial Revolution. To an extent never suspected in agrarian societies, the machine age demonstrated a capacity to create sufficient wealth and eliminate, for the first time, great masses of poverty. This ideology of progress was buttressed by the narcissism implicit in Darwin's theory of evolution, which suggested that the human species was the most apt of all the species on the planet, due to its capacity to adapt itself to any natural habitat and always to achieve progress.

Industrial evolutionism engendered the conviction that any society can create science, technology and industry, thereby progressing without limits. Just as the *Australopithecus* progressed to become a tool-wielding man, then to *Homo erectus*, who created fire, and later to *Homo sapiens*, who created language and culture, rural societies can progress from being agricultural producers to industrialised societies. Finally they become postindustrial societies endowed with intensive knowledge and perpetual well-being.

The certainty of the nineteenth-century evolutionary industrial prediction was reinforced during the twentieth century by a tide of inventions that gave rise to epistemological optimism. This leads man to believe that all problems can be solved by science and technology. If a problem appears to have no solution, that is a momentary impression, because it is practically certain that the technology to solve it will be invented. Therefore, there is no doubt about scientific material progress and its relation to human happiness.

In the context of this ideology of happiness based on material progress, already in the eighteenth century Adam Smith described the stages required in order to achieve the wealth of nations. He explained how the societies of hunter-gatherers could evolve towards pastoral and agricultural communities, to finally become manufacturing and trading societies. The thinking of Karl Marx, another great ideologist of human happiness through material progress, followed Smith's evolutionist path. For Marx, humanity's material progress is attained by passing from feudalism to capitalism and from there to communism, thus ending history and spawning perpetual happiness. In this, Marx partially coincided with the neoliberal Francis Fukuyama, who considers that today history has ended with the triumph of global capitalism.

One of the great modern promoters of the myth of development was Walter Rostow, a professor at the Massachusetts Institute of Technology. In 1960, he fascinated all the technocracies with his famous book on the stages of economic growth. According to Rostow, countries evolve from a traditional society, through stages of accumulation and take-off, to reach the final stage of mass consumption, which he says is nothing less than development. The natural environment has no importance in this process. It is just another raw material to consume on the march to progress and happiness. After Rostow, all the technocrats were convinced that they could achieve development. They only needed to know how to apply the correct theories and policies, create value added, accumulate, take off and indulge in mass consumption. The idea was to reproduce in the shortest possible historical time the development processes of Europe and the United States. Since the 1960s, we have witnessed many 'take-offs', but few cases of national development. Twenty years ago, it was said that Brazil was taking off, that it was one of the future world powers. Then, some years ago, Mexico was in fashion, then India. This was followed by the vogue of the 'emerging countries' of Asia. Today the only take-off in fashion is that of China, a country with 1.2 billion inhabitants, where only 300 million have a standard of living that would permit them to be consumers in the global economy.

 The fact is that in the last thirty years only two small countries, South Korea and Taiwan, have managed to progress from agricultural societies to technologically advanced industrialised societies. They have conquered their generalised poverty and raised living standards to create a predominant middle class. However, this was done with democratic, cultural, scientific and social levels far below those of Europe and the United States.

Two other territories, termed by the development gurus newly industrialised countries (NICs), Hong Kong and Singapore, which have also approached the living standards of the developed capitalist democracies, are not nation-states but small city-states. Their development did not confront the enormous problems entailed in raising the living standards of vast territories with an ongoing and unstoppable urban population explosion, as is the case with the majority of underdeveloped countries. Today, when underdevelopment is the characteristic of most of the planet's nation-states, and when environmental deterioration is caused by the progress (mass consumption) of a minority of industrialised countries, the words of Rabindranath Tagore seem particularly apt: 'progress for whom … progress toward what?'

During the Cold War, the myth of development was expressed basically in two rival models: the communist and the capitalist. Both were in environmental terms unsustainable. Now that communism has collapsed because it tried to replace the market with centralised planning of scarcity, a type of global capitalism is emerging that has gone to the extreme of converting the market into a kind of supreme natural law, ethically neutral, like the law of gravity. As such it pays absolutely no attention to social and environmental considerations, and expects everyone to submit passively to this. This model, at present the only one to express the myth of development, attempts through globalisation to reproduce modern capitalist societies in the majority of the underdeveloped countries. However, contrary to expectations, the prosperity explosion expected in the wake of capitalism's triumph is not becoming reality. History has not ended. On the contrary, it is becoming complicated, because what is actually happening is the exclusion of major sectors of the world's population from the global economy.

However, since the myth of development has nearly religious connotations of hope and salvation from poverty, it remains untouched by the experience of the last forty years, which demonstrates so unequivocally the utter lack of development of the majority of countries. The mythical nature of development leads the politicians of poor societies to continue insisting on 'closing the gap' that separates them from the capitalist industrialised societies – closing it by attempting to reproduce consumer patterns that cannot be financed or sustained environmentally.

The myth of development has impregnated our civilisation to such an extent as to inspire splendid, headstrong international stances like the United Nations proclamation of the 'right to development', that is, the right of all underdeveloped countries to have living standards and consumption patterns like those of the industrialised states. The recognition of this right in United Nations declarations bears no relation to its real chances of becoming effective. Besides, its hypothetical achievement at present Western consumption levels would cause an environmental catastrophe on the planet.[1]

Outside UN conference rooms and in the real world, there are countries that are simply incapable of 'closing the gap' with the industrialised countries, even with the most unfettered economic policies and the most abundant and cajoling international co-operation. The so-called developing world is rife with countries that have no modern capitalist class, and no scientific and technical personnel capable of using the liberalisation of the economy and foreign aid to modernise the economy and make it competitive on a global scale. In these countries, the explosive growth of the urban population is producing very high levels of poverty and unemployment, as well as social, religious and ethnic divisions, accompanied by deficits in food, energy and water, which are the minimum resources required for the existence of an organised society, a nation-state.

To add to this worrisome situation, in the real world, as we saw in Chapter 3, international relations are Darwinian. The global economy increasingly demands products and services with a high technological content, while the underdeveloped economies are still trapped, like inapt species, in barely processed production with no technological innovation. Moreover, the new power of the transnational corporations keeps pushing globalisation by opening up markets with environmentally unsustainable consumption patterns, making the poor countries import more. This increases their foreign debt, without helping these countries to modernise and be able to compete in the global economy of the future.

International aid, the daughter of the myth of development, is paradoxically the clearest testimony of non-development. During nearly half a century, the United Nations, industrialised powers, specialised agencies, international financial organisations, non-governmental organisations and humanitarian institutions have tried uncounted policies, strategies, programmes, and development projects, transferring billions of dollars in credits, technical assistance, equipment and donations. Part of this enormous mass of resources has been recycled through tied aid. Another part was lost in the corridors of corruption, and only a modest stream of this torrent of resources has been applied to alleviate poverty.[2]

The crude reality is that today nobody knows how to reach El Dorado. The rich are getting richer and the poor poorer, in all countries. The combined income of some 300 individual billionaires is equal to the total revenues of 2.7 billion persons who represent 45 per cent of the world's population. The individuals who have the means to consume the products and services of the global economy number only 1.8 billion. The remaining 4 billion-plus are left window-shopping. In nearly a hundred poor countries, real per capita income has not increased in fifteen years.

If the present trends continue, and nothing indicates that they are going to change, in the year 2020, the world population will reach 8 billion, of whom some 6.6 billion will live in the underdeveloped world, where there will be 3 billion poor, plus 840 million who are starving and hundreds of millions who are unemployed or, at best, underemployed. In addition, 2.5 billion will not have adequate housing and 2 billion will have no access to clean water or a commercial energy supply. The overwhelming majority of these marginal inhabitants will live in more than 550 cities with populations over 1 million and some twenty megalopolises of more than 10 million inhabitants. These cities will be chaotic, polluted, full of unemployed workers and plagued by delinquency. The harbingers of this nightmare are already on view in Lima, Sao Paulo, Bogota, Lagos, Cairo, Nairobi, Dhaka or New Delhi.

Even though these trends, which can so easily be confirmed by a tour around the so-called developing world, have been utterly clear throughout the 1990s, the World Bank, the IMF and many specialised economic circles continue to predict a rosy future for the 'developing' or 'emerging' world. Possessed by the myth of development, they habitually predict El Dorados, identify miracle countries, future economic powers. Today it is China, and perhaps tomorrow it will be Turkey. They describe precarious economic recoveries without technological modernisation, like those of Peru, Argentina or Ghana, as 'miracles'. Then come the debacles, but by that time they are already predicting other national miracles.[3]

In their analyses, they hint that the growth of China or of the Asian NICs can be copied in the other countries, even in Africa. They are convinced that the present globalisation process will create world prosperity. They comment and announce take-offs, economic miracles, praise painful refinancing of foreign debts, wax ecstatic over the high growth of the GNP. However, they say precious little about the chaotic world behaviour patterns that are emerging.

The IMF and the World Bank failed to foresee the main financial debacles of the twentieth century: the debt crisis in Latin America, Mexico's insolvency, the bankruptcy of the Asian countries, Brazil and Russia. With obsessive optimism, they supported nothing less than a plan to convert the USSR to a market economy in five hundred days! The result has been a cataclysm of social exclusion and the birth of the world's first model of a klepto-market economy. The ideological unreality of the experts of the World Bank and the IMF does not allow them to see even the poverty and delinquency next door to their own offices in Washington, a poverty that has been caused, precisely, by a model very similar to the one they are recommending for everyone all over the world. The truth of the matter is that, in order to have some idea of what is going on in the world, one should travel more and spend less time reading the reports of the World Bank and the IMF.[4]

In the streets of Africa, Asia and Latin America, it is easy to confirm that the majority of the population is leaving rural life but is not becoming a broad middle class, conscious of its civic duties and of its democratic rights, as happened in Europe and the United States. Instead, a large part of the population is living in poverty and with rudimentary civic behaviour. In most countries, it is not democratic capitalism, coupled with modernised production, that is arising. Rather, what has emerged is backward, rustic capitalism, in low-intensity, even grotesque democracies, anchored in primary, barely processed production. Moreover, there is no rule of law, institutions are not respected, and even less is civil society.

The gurus of the myth of development, who measure everything, have virtually a quantitative vision of the world. They pay no heed to qualitative historical and cultural processes, to the non-linear progress of society, to the ethical point of view, and they even disregard the environmental impacts. They misconstrue economic growth as the development of a capitalist modernity that does not exist in the poor countries. With such a perspective, they only perceive the economic epi-phenomena such as GNP growth, export performance or the behaviour of the stock market; they do not notice the profound qualitative cultural, social, environmental and structural dysfunctions that prefigure the non-viability of the underdeveloped quasi nation-states in the new millennium.

In order for them to become aware of that, they need to take into account the degree of scientific–technological integration that these countries have attained in the global economy. In addition, they must examine the physical and social equilibrium between the growth rate of the urban population, on the one hand, and the future supply of the indispensable physical resources needed for national cohesion and a civilised life, such as food, energy and water, on the other.[5] Only with this sort of vision will it be possible to detect the 'viruses of non-viability' affecting many of the presently misnamed 'developing' countries.

Non-viable national economies (NNEs)

In our time, the nations' future depends increasingly on know-ledge and information, that is to say, on the number of scientists and engineers, on the expenditures on scientific and technological research and on the production of software. The underdeveloped countries that constitute 75 per cent of humanity (4.5 billion inhabitants) possess only 7 per cent of the world total of scientists and engineers, spend less than 2 per cent of the total world invest-ment on scientific research and development, and produce only 3 per cent of the software. These estimates of backwardness may seem generous, because in fact half of this laughable arsenal of science and technology is concentrated in a handful of countries, basically Singapore, Hong Kong, Malaysia, Taiwan, China, India and Brazil. All the rest live in the most absolute scientific and tech-nological destitution, a situation that will gradually exclude them entirely from the global economy, where the demand continually increases for sophisticated manufactures and services.[6]

The main disease, therefore, that is infecting, with increasing virulence, the vast majority of the misnamed developing countries, is scientific and technological poverty. Indeed, today the world demand for products and services with a high technological content is growing by 15 per cent per year, while the growth rate for raw materials is under 3 per cent and for semi-processed goods, 4 per cent a year (see pages 100–105). According to the World Bank, real raw material prices, which have already fallen below the levels of the Great Depression of 1932, will continue to decline well into the twenty-first century.

The problem is compounded by the fact that not only will raw material prices remain unstable and hardly remunerative, but the same will be true of prices of manufactured goods with low or medium technological content. Recent studies by UNCTAD confirm that the prices of manufactured goods such as textiles, clothing, wood products, chemicals, machinery and transportation equipment exported by the African, Asian and Latin American countries have fallen by 1 per cent a year since 1970. This perverse downward trend of prices is not unlike that of raw materials.[7]

When the virus of scientific and technological poverty colludes with another non-viability virus, such as demographic explosion, non-development is virtually inevitable. This occurs because the meagre export income that many countries will receive in future from raw materials, agricultural products, and goods with low technological content will in no way be sufficient to create employment and satisfy the needs of the burgeoning urban populations in the underdeveloped cities.

Practically all the countries with low-technological-content exports will have doubled their population by the year 2020 or thereabouts. This viral combination of scarcely processed exports and demographic explosion is a leading producer of poverty. Today nearly half of the total population of Africa lives in poverty. This is also true of 40 per cent of the inhabitants of the large and populous countries of southern Asia, like India, Bangladesh, Pakistan and the Philippines. In Latin America, poverty affects nearly 38 per cent of the population, and is especially concentrated in the Central American and some of the Andean countries, such as Ecuador, Peru and Bolivia. If the underdeveloped countries are unable to modernise their exports, adding greater technological content, and if they fail to reduce their birth rates, the poverty that now affects 1.3 billion people will affect nearly 3 billion by the year 2020.

As a result of their inability to obtain resources from the world market for their growing populations, countries do not develop. Instead, they acquire the characteristics of non-viable national economies (NNEs). All the so-called 'developing' countries have been forced to survive during nearly the entire twentieth century on international aid, official loans and credits from private institutions, continually falling into insolvency and national bankruptcy. At present, they are surviving by dint of privatisations and with a great deal of volatile capital, generated by speculation in the global financial market.

One of the most illustrative examples of the characteristics of NNEs is the economic history of the greater part of the Latin American countries. During an entire century, the short-lived booms of the region were due principally to two factors: the temporary rise in price of some primary products and the coincidence of these price peaks with a period of abundant foreign capital and credit. This facilitated loans and investments, which later ceased.

In 1920, the spate of liquidity around the world resulting from German war reparations and from the abundance of US capital, produced various 'miracles' in Latin America, which ended in the crash of 1930. The Second World War, the reconstruction of Europe and the war in Korea all helped to raise the prices of primary products and increase the supply of foreign investments in Latin America. This brought a new period of expansion which ended with crises and new recessions during the final years of the sixties. In the seventies, another wave of liquidity and abundant capital supply from recycled petro-dollars kept the economies afloat with loans from private banks. This whirlwind ended in the catastrophic debt crisis and the consequent insolvency of the Latin American economies. At the end of the twentieth century, global deregulation of the financial markets and the emergence of a global capitalist casino have again permitted miraculous recoveries, with short-term investments based on speculation and privatisations. When the drunken spree with the present financial liquidity has passed, the technologically backward economies of most Latin American and other underdeveloped countries will again be depressed under the weight of their unchecked urban expansion.[8]

The NNEs are a fundamental cause of the social disintegration, and eventual collapse, of the underdeveloped nation-states. In the majority of cases, none the less, economic non-viability does not necessarily degenerate into a terminal crisis of the nation-state, like a fatal illness in living beings. Non-viable economies can continue to stay alive for decades in a stabilised condition (non-development), undergoing the typical historical alternation of crises and miraculous recoveries, without eliminating the viruses of non-viability or diminishing national poverty to any noticeable extent.

This stabilised stage of non-viability, in which poverty does not decrease significantly nor does the nation-state collapse, is supported by temporary bonanzas that fail to modernise the economy. These bonanzas may be triggered in various ways: by a momentary rise in the price of some primary export product, by new, slightly processed exports (such as those assembled in the country from foreign parts), by drug traffickers' investments, by moneys sent home from citizens living abroad, by short-term speculative investments, by privatisations or by financial rescues at the hands of the IMF or the World Bank. These countries may also be refloated by a consortium of industrialised powers who do not want certain nations to suffer a socio-political disintegration, which might affect their own national interests.

Within the context of this sort of stabilised economic non-viability, it is possible for the GNP of an underdeveloped country to register miraculous, albeit short-lived, periods of growth, thus confusing and elating observers, who believe that the country has finally taken off as a newly industrialised country (NIC), that it is becoming a new 'tiger' who will prowl the global market with the best of them. A tourist in a luxury hotel situated in one of the rare high-income ghettos of an underdeveloped country could report having found good restaurants, new automobiles, shopping centres full of foreign consumer products and artefacts. Upon departing, he would be convinced that the country in question is well on the way to development. This illusion is further heightened by some of the national and international press, with a fondness for discovering and highlighting 'miraculous' economic stories.

These cases of economic growth do not represent the sustained development of a modern, competitive capitalist economy, because the technological content of their contribution in the global economy has not improved. They are not the outcome of a critical mass of productive national and transnational investments that create jobs and spur a modernisation process. On the contrary, they are the offspring of extremely volatile global speculative financial flows that could easily be withdrawn.

Ultimately, when these cycles of stability and precarious growth have ended, the only thing that develops with the passing years in countries that are stabilised in non-viability is the growth of their low-income population. The best way to recognise whether a country shows symptoms of non-viability is not to observe the temporary spates of GNP growth, but to see whether its scientific and technological research is increasing, whether its exports are being modernised, whether personal incomes are rising, and whether national poverty is starting to decrease significantly and steadily, year after year.

A UNDP study (1997) on the average growth of personal incomes over 34 years (1960–94) concluded that only four countries of Latin America (Chile, Colombia, Brazil and Ecuador) have achieved an average growth of per capita income of just 2.5 per cent. During those thirty-four years, the remaining countries, and in particular Peru, Bolivia, Central America and Haiti, have registered an average of zero growth in the personal per capita income. That is to say, the only thing that has grown is the population. These are truly dysfunctional countries with strong signs of economic non-viability. In Africa, the situation is still more disastrous. Practically the entire continent has registered an average income per capita equal to zero. And in many countries incomes have lost value. The same process has affected a majority of the South Asian countries. The only countries of that region where poverty has declined markedly are Hong Kong, Singapore, South Korea, Taiwan and China. There, during the last thirty-four years before the crisis, personal incomes registered an uninterrupted average yearly growth of nearly 7 per cent.

By the end of the twentieth century, the process of economic non-viability had left most of the underdeveloped countries with nearly 40 per cent of their population in a deplorable state of human development and with their middle classes struggling to avoid slipping into poverty. Only a minute minority enjoys patterns of consumption and living standards similar to those of the industrialised countries. In the countries with non-viable econ-omies, the majority of the population lives in a hell, a small middle class lives in purgatory, and only a handful enjoy the paradise of a consumer economy, with instantaneous gratification.

Today, there are already countries in Africa and Asia that display all the characteristics of NNEs, and are being left on the sidelines of globalisation, through the process of selection applied by the market and technology. In this dire situation are to be found the greater part of the countries of sub-Saharan Africa, such as Angola, Burkina Faso, Burundi, Central African Republic, Chad, Congo (ex-Zaire), Côte d'Ivoire, Equatorial Guinea, Ethiopia, Gambia, Guinea-Bissau, Kenya, Lesotho, Liberia, Madagascar, Malawi, Mauritania, Mozambique, Niger, Rwanda, Senegal, Sierra Leone, Somalia, Sudan, Tanzania, Togo, Uganda, Zambia and Zimbabwe. In Asia, Afghanistan, Bangladesh, Bhutan, Cambodia, Myanmar (Burma), Nepal and Yemen fall into the same category. In Latin America, the only economy with definite signs of non-viability so far is Haiti. However, as will be seen later, the symptoms of non-viability are also beginning to appear in the economies of Central America and the Andean region.

Of all the above-mentioned African and Asian countries, not one of their economies has a trace of technological modernisation in its export products; they hardly receive any productive transnational investment. Almost all are mono-producers or depend on the export of a few primary products with unstable prices. Their population growth rate is an extremely high 3.1 per cent per year. By the year 2025, they will have doubled their total of inhabitants, while their food production diminishes. Thus the populations have less food security every day. Food imports represent between 25 and 80 per cent of their total imports. They also lack energy security. Compared with the yearly energy consumption in an NIC of 1,000 kilos of petroleum per capita, these countries use only 120 kilos. Due to this lack of access to energy, the population resorts to firewood, thereby causing rapid deforestation, which erodes the soil and reduces food production. Their imports of energy average about 22 per cent of their total imports. The same desperate situation applies to their water security. More than 70 per cent of their population lacks drinking water and sanitation. In all these countries, almost 53 per cent of the population live in poverty; therefore, these countries cannot build an economy based on a national market.[9]

Alongside this group of African and Asian countries, there is another group of underdeveloped countries in Latin America, Asia and the Middle East whose economies are incubating similar symptoms of non-viability. This second group includes countries like Algeria, Bolivia, the Dominican Republic, Ecuador, Egypt, El Salvador, Guatemala, Guyana, Honduras, India, Jordan, Morocco, Nicaragua, Pakistan, Paraguay, the Philippines, Peru, Sri Lanka, Syria and Tunisia.

These economies export mainly raw materials and manufactured products with low technological content, for which the demand is not growing appreciably. The prices for these products are barely profitable and in any case are insufficient to provide the resources necessary to overcome the poverty and unemployment resulting from their urban population explosion. Like the African and Asian countries in the first group, these countries do not receive the critical mass of productive transnational investment that would serve to modernise their exports, giving them a start towards a competitive advantage in the global economy.

The countries of this second group are also characterised by a very high yearly growth rate of the urban population, above 2.5 per cent. All of these countries will double their population by the year 2025. Many of them will have various cities that pass 1 or 2 million inhabitants and, by the year 2010, some of them will have megalopolises of more than 10 million inhabitants, as will be the case with Manila, Lima, Cairo, Dhaka and Karachi.[10]

A majority of these countries are beginning to lose ground in terms of food security; many already depend on international food aid to cover the calorie deficit of their populations. In this situation, Jordan, Peru, Bolivia, Egypt, Pakistan and all of Central America are to be found. These countries are also starting to lose their ability to cover their energy requirements. A majority of them will import increasing amounts of petroleum, even though their per capita consumption of energy is extremely modest. Only Ecuador and Algeria still export petroleum and gas, but their future energy security too is under threat, because of the high growth rates of their populations. In addition, all of these countries are beginning to suffer a serious shortage of water, as the result of their urban population explosion. While the poverty in these countries is less severe than in the African and Asian countries of the first group, it is still very great because it involves between 30 and 40 per cent of the population. As a consequence, since a large part of the population is poor, with incomes of less than one dollar a day, these countries cannot create a genuine national market economy.[11]

Both the first group and the second group have in common an export structure that is technologically dysfunctional with the global economy. These countries continue to export virtually the same primary agricultural and mineral products that were in vogue in the nineteenth and the first half of the twentieth century. Their archaic integration corresponds to the first capitalist industrial revolution, typically intensive in its use of raw materials and abundant unskilled labour. This type of exportation is now totally out of touch with the trends of the new global capitalist economy.

Another similarity between these two groups is that their participation in world trade has diminished. For example, between 1982 and 1995, Algeria, Bolivia, Cuba, the Dominican Republic, Egypt, El Salvador, Guatemala, Haiti, Kenya, Nicaragua, Nigeria, Peru, and Yemen did not increase or diversify their exports in any significant way so as to acquire more revenues. In contrast, during the same period the economies of such countries as Colombia, Mexico and India doubled their exports; Chile, Malaysia, Argentina and Singapore tripled their exports, and Thailand and China quintupled theirs.[12] This contrast reveals important differences between those underdeveloped economies that have been able to diversify and that appear to have a chance of entering the global capitalist economy of the twenty-first century and those underdeveloped economies whose exports have not diversified, and have not grown at an adequate pace to extricate them from poverty.

My description of these two groups of countries of Africa, Asia, Central America, the Andean Region and the Middle East as incubating symptoms of economic non-viability does not mean that I judge that the future is necessarily rosy for the underdeveloped countries not included in these two groups. Nor does it mean that their manifest destiny is to be like Taiwan, South Korea, Hong Kong or Singapore. For example, Argentina, Chile, Colombia, Costa Rica, Mexico, Uruguay and Venezuela have increased their exports, have reduced their demographic growth rates and have established food, energy and water security. However, the goods these countries export are still principally primary products, or manufactures with far less technological content and global competitive advantages than the exports of South Korea, Malaysia, Singapore, Hong Kong or Taiwan.

Argentina and Venezuela alone may be sure of being more or less integrated into the global economy in the next twenty years, thanks to their positions as important world exporters of petroleum and cereals, two extremely strategic products for the future in view of the world urban population explosion. In contrast, Chile, Colombia, Costa Rica, Mexico and Uruguay are still a long way from becoming NICs, and if they do not undertake a decisive modernisation of their exports to give them greater technological content, they probably will also suffer exclusion from the global market in the coming years.

There is no guarantee, either, that such mega-countries as China, India and Brazil will manage to reduce their enormous poverty and raise their paltry human development levels to a standard of living at least near to that of the present-day industrialised countries. While it is true that these countries have very large economies, they also have colossal social and environmental problems. These giants are suffering an urban explosion process with burgeoning consumption patterns that place enormous pressure on the environment and great demands on the supply of vital resources, like food, energy and water. In these three countries will be found the principal megalopolises of the twenty-first century. This mega-urbanisation will transform the economies of China, Brazil and India into gargantuan consumers of energy, food and water. How will these countries resolve their physical and social imbalances? No one can tell. The only certainty is that it will not be easy.

Non-development

Two hundred years after the emergence of the modern capitalist nation-state, and more than forty years into the reign of the myth of development, reality shows that the rule is the non-development of at least 130 countries and the exception is 4 newly industrialised countries (NICs) – Taiwan, South Korea, Singapore and Hong Kong. If we analyse in more detail this historic result, we must admit that during practically an entire half-century only two small nation-states – South Korea and Taiwan – and two city-states – Singapore and Hong Kong – have been able to develop into modern capitalist countries. In total the four represent a mere 2 per cent of the population of the misnamed 'developing world'.[13]

These are the only cases that could be said to have undergone a process similar to that experienced 150 years ago by the present-day capitalist industrialised powers. That is, they have achieved a productive technological transformation and a sizeable conversion of the poor into middle-class citizens. Nevertheless, these NICs still do not enjoy the high incomes or the scientific and cultural development, much less the democratic institutions and civil society, characteristic of Europe or the United States.

Because the world strategic context has changed, it would be impossible to reproduce the emergence of this small archipelago of four almost-developed capitalist economies in the vast ocean of world underdevelopment made up of the poor, low-income populations of Africa, Asia and Latin America. During the seventies, the Cold War meant war in Asia. The threat of Sino-Soviet expansion was tangible in Vietnam. The United States and Japan were obliged to reinforce the economies of Taiwan, South Korea, Singapore and Hong Kong as bastions to contain Communism and to avoid a domino effect in case Southeast Asia were lost. For that reason, they flooded them with abundant international investments and credits and allowed them something that is today considered taboo to create state-supported capitalism, protecting and orienting private enterprises toward exports. These four countries also received large investments from Japan, a country that, under the protection of the United States, was spared the heavy costs of an arms race.

Today the situation is radically different. The end of the Cold

War and of the Communist threat for many countries also put an end to their strategic opportunities of attracting capital investment. Instead, an ultra-liberal form of global capitalism bars the state from a managerial role in support of comparative advantages for national capitalism. On the other hand, the real prices of raw materials and of manufactured goods with low technological content, such as those that the Southeast Asian NICs were exporting initially, have deteriorated compared to their prices in the sixties. This means that it is no longer possible to be competitive with such manufactures, as the NICs were in the seventies.

Besides, it is highly unlikely that the countries of Latin America, Asia and Africa will receive a tidal wave of foreign capital for investment in productive export activities, as did the Asian NICs twenty years ago. In today's world financial market, very little capital is available for productive, job-creating investment, and most of that goes to the industrialised countries themselves (see Chapter 3). The same foreign investment that was used to create factories and a huge exporting platform in Asia some decades ago has now become so speculative that it even drove into bankruptcy an NIC, South Korea, along with other emerging Asian countries such as Thailand and Indonesia.[14]

Despite all, the conviction remains, among many leaders of the Group of Seven and among the international technocrats of the IMF, the World Bank and the WTO, that a modern capitalist process can emerge in the backward and poor countries by merely liberalising and privatising their non-viable economies. But Peru, Egypt, Morocco, Algeria, Thailand, Philippines, Indonesia, Kenya, Nigeria or Pakistan may liberalise, deregulate and privatise their economies, but this alone, without technological modernisation of their exports and without solid democratic institutions, will never suffice to create a genuine economy and a modern capitalist democracy. It is tantamount to buying a computer without the necessary software. With governments that are elected but are not democratic, that do not accept supervision by the judiciary and an autonomous parliament, the result will be a rustic, even brutal, capitalism, devoid of democracy, fuelled by speculative capital, plagued by political favours and klepto-capitalists. The GNP may grow, but national prosperity will not.

The only way to protect the poor countries from becoming economically non-viable would be rapidly to modernise their production and to enable them to start to export manufactures and services with increasing amounts of technological content. They have to free themselves from the trap embodied in the exclusive export of minerals, agricultural products, woods, hides, beverages, textiles, and to trade in other, more sophisticated products, such as electronic equipment, semiconductors, biotechnology, pharmaceuticals, petrochemicals and, above all, software and parts for the transnational telecommunications, transport and aerospace industries. They need to invest, as well, in more competitive national services, modern infrastructure and, particularly, in scientific and technological research and development.

This modernisation cannot be brought about with the woefully insufficient national capital and the non-existent domestic scientific–technological resources that these countries possess. Experts estimate that this modernisation would require a critical mass of transnational investments and technologies of at least US\$300 billion per year.[15] As noted above, the chances of receiving this critical mass of direct investment from transnational corporations to modernise the backward economies are virtually non-existent (see pages 48 and 90). But even in the hypothetical case that this critical mass of transnational investment were to be directed to the underdeveloped countries, and were to begin the longed-for modernisation, the new industrial processes would have to use labour-saving modern technologies. They would be unable to create employment for the nearly 30 million unskilled workers who are seeking jobs in the large cities of Africa, Asia and Latin America.

In fact, if the poor countries did industrialise and reproduced the production and consumption patterns of the present growth model, the environmental cost would be disastrous. Lands, rivers, seas and lakes would be heavily polluted and the gases emitted would further affect the climate of the entire planet. An example of such environmental damage is provided by the Asian and Pacific countries that received the bulk of the few productive investments in the underdeveloped world between 1970 and 1990. In the year 2020, these countries alone will produce one third of the world's emissions of gases into the atmosphere. What would happen if all the poor countries, including China and India, were to adopt the present models of production and consumption?

In fact, however, this could hardly happen. The famous El Dorado of development, by means of an export-based economy, has inescapable limits. Such models only succeed when the number of world players is limited, as was the case for South Korea, Taiwan or Hong Kong. In contrast, if all the world, including mega-countries like India, China and Brazil, are eager to play the game of exporting tigers, no one wins. How could each country in the world increase its annual exports by nearly 20 per cent when the world economy grows by only 3 per cent a year?

In the twenty-first century, there will be frequent cases of states that collapse in deep economic crises, and then recover to become stabilised in a state of non-development. Their alternative may be to implode into violence, as has already happened in parts of Latin America, Africa and Asia.

CHAPTER 5

WORLDWIDE DEPREDATION

depredation
robbing, plunder
raging waste
(destroy)

Treatment as a different species

During the entire second half of the twentieth century, we grew accustomed to hearing that development could be achieved, violence would cease and history would be ended, all by changing our economic and social systems or structures. As we begin the third millennium of our era, we must emphatically deny the truth of such a belief. If the systems and structures continue to produce poverty and violence, it is because they are a reflection of the predatory nature of our own species. The history of *homo sapiens*, studied without narcissism, leads to the inevitable conclusion that, ultimately, man is still a predator of his own species. This tendency has been constant, in greater or lesser degree, for the last ten thousand years of civilisation. Such depredation has occurred under all sorts of socio-political systems, from slavery, feudalism, absolutism, colonialism, and Nazi-Fascist and Communist totalitarianisms, even to capitalist democracy. What is more, all cultures and civilisations have rationalised and glorified the maximum exponent of human depredation: war. This predatory activity has been considered a legitimate means of bringing to fruition the ambitions of the tribe, the clan or the nation-state. Depredation among humans has been and still is a historical constant, one that has not yet been tamed by any great ethical or religious current.

Depredation as a constant in human history is rooted in an innate tendency of *homo sapiens* to consider his own group (family, tribe, clan, ethnic group, nation, religion or culture) different and superior to other human groups, thereby dividing the human species between *them* and *us*. Doing this means that we do not share our humanity with other human beings who belong to different ethnic groups, creeds or cultures, and we go so far as to treat them as a different species, thus engendering the concept of 'the enemy'. This predatory rationalisation is called by the German ethologist Eibl Eibesfeldt pseudo-speciation.[1] Humans have an instinctive tendency to pseudo-speciate other humans, treating them as though they were a different species.

Examples of the abasement of humanity are common throughout the history of man, and they continue to occur today, in the form of social exclusion, frequent violations of human rights, terrorism, or the civil wars of extermination and genocide that occur in different regions of the world. Today, human groups that feel different and superior because they dominate new technologies, treat as virtually another species those who are less advanced. Conversely, human groups with traditional cultures who feel offended by so much modernity consider the more technologically advanced groups as a satanic product to be violently opposed. From the tribe to the modern nation, humans continue to mistrust other humans with whom they feel no affinity.

Starting with the Enlightenment, Western rationalism maintained that this instinctive predatory tendency of man could be controlled by constructing societies based on justice. From that time onwards, the quest for happiness in the West would become the art of organising just societies. The first attempt was the French Revolution, which gave birth to the modern concepts of revolution and of political and social engineering. The last of these would later inspire the Bolshevik totalitarian revolution and many other national revolutions, as well as the organisation of the modern nation-states, running the gamut from liberal to totalitarian regimes. Paradoxically, as soon as the ideas of revolution and political and social engineering were born, the human predatory tendency that they were meant to control became even more vicious. This was due to the fact that the political and social

engineers themselves saw each other as mortal enemies, since they
espoused rival projects for happiness.

For the above reasons, our Western civilisation – the most technologically advanced of them all and the cradle of human rights,
democracy, and the idea of progress – has practised human depredation to a degree unequalled by any other civilisation. In the
name of a just society, two world wars were unleashed in the West,
wherein the scientific and industrial power of the countries
involved was transformed into an infernal machine wreaking anni-
hilation. The First World War mobilised 65 million soldiers, 9
million of whom perished under a hurricane of iron and shrapnel,
produced by the modernisation of the firearms inherited from the
turn of the century. The Second World War, with 92 million
combatants, produced 50 million victims, both military and
civilian. This was the overall result of massive aerial bombings,
automatic weapons, gas chambers and two atomic bombs. The
machinery for total destruction was thus improved, making cities
and civilian populations also the target of the substantial progress in
lethal technology. In all, treatment as a different species, cloaked in
just causes and patriotic fervour, wiped out 150 million people
during the first half of the twentieth century.[2]

In the second half of the century, treatment as a different species
emerged as a strengthened ideology. In addition to the promotion
of patriotic duty to defend the national interest against the enemy,
crusades were launched to bring about the triumph of either the
capitalist or the communist ideology, these being the outstanding
exponents of projects to establish a just society and to reach
material prosperity. A series of low-intensity conflicts broke out,
controlled by the United States and by the USSR as representatives of these rival projects for happiness. These did not spread into
world conflagrations, for fear of nuclear retaliation that would
mutually destroy the two superpowers and possibly extinguish a
great part of life on the planet. Consequently, the treatment as a

different species that had triggered two world wars was brought under control, for the first time in the history of the world, not by justice or by ethical considerations, but by 6,000 nuclear missiles, each one a hundred times more powerful than the bombs dropped on Nagasaki and Hiroshima. The fear of nuclear arms raised for the first time the spectre of the extinction of the species and made us see ourselves as one single human race. For the first time in hundreds of years, this fear prevented our recourse to war as a means of imposing just societies and national interests.

The ideological war between capitalism and communism during the second half of the twentieth century was not a conflict between totally different ideologies. It was, rather, a civil war between two extreme viewpoints of the same Western ideology: the search for happiness through the material progress disseminated by the Industrial Revolution. Both capitalism and communism are products of the Industrial Revolution's manufactures. The first represents the individualistic and democratic approach based on the market and inspired by Anglo-Saxon thinking; the second embodies the collectivist and authoritarian approach stemming from German political philosophy: two predatory interpretations of the same ideology of material progress. Each of them treated the other side as a different species, thereby denying their mutual humanity. Surrounded by nuclear missiles and bloated with propaganda and misinformation on an unprecedented scale, capitalism and communism accused each other of being inhuman systems, while presenting themselves as champions of a higher, more ethical world order, one that definitely excluded human depredation.

 There is no doubt that totalitarian communism was by far the more dehumanised of the two, because it did not even recognise the freedom of the individual to question its ideological project for happiness. Real socialism never admitted that civil society and the law might stand above the interests of the Communist Party hierarchy. Nothing shows more clearly the greater degree of inhumanity in communism than the imperialist wars waged by both sides before the end of the Cold War. In the case of Afghanistan, despite the fact that the majority of the Soviet people opposed the war, they never dared to express their disagreement for fear of repression and due to the total suppression of public opinion in the USSR. A similar situation never existed in the capitalist democracies, where strong currents of public opinion criticised the United States' intervention in Vietnam, eventually forcing the US to withdraw.

The Soviet disinformation published during the ideological civil war kept the developed capitalist societies paralysed with fright during nearly forty years, leading them to believe that communism was as efficient as capitalism in achieving material well-being. This Soviet propaganda mitigated capitalism's predatory attitude in several ways. It forced the capitalist democracies to grant independence to the colonial peoples of Asia and Africa, to establish international development aid, and also to introduce social reforms that humanised the capitalist democracies. In the long run, communism's totalitarian depredation began to produce its own technological backwardness and an enormous scarcity of material goods, finally convincing its own leaders that the system was non-viable. Thus, to the astonishment of the capitalist democracies, the collectivist side of the Western ideology of progress collapsed, ironically, due to its inability to produce that same progress.

The triumph of the capitalist side of the ideology of progress has not, by any stretch of the imagination, meant that we are entering upon a new ethical order and have overcome depredation. On the contrary, we are today facing world hegemony in its most predatory, most individualistic version. This brutal version of capitalism pursues happiness as never before, to be achieved by the highest degree of material accumulation. It spreads environmentally unsustainable patterns of consumption, and uses the market and technology to plunder persons, enterprises and nations. Now that the fear of communism has faded, the only interest lies in higher profits, quick and easy money, with no thought for the environmental and social costs.

Stands in favour of democracy, freedom and human rights, which are the most valuable aspects of the capitalist version of the Western ideology of progress, would seem to have been merely an anti-communist tactic. In the post-Cold War era, all too many examples can be found of a global capitalism that gives up its defence of democracy, freedom and human rights. As examples could be mentioned the general indifference in the face of the genocides in Africa, the benevolence to the ayatollahs that liquidated many of their opponents who had found asylum in Europe, the unholy alliance between the West and the oppressive sheikhs of the Gulf, the ambivalent response to electoral fraud in Peru and the tolerance of the corrupt and authoritarian, low-intensity democracies of Africa, Asia, Latin America, the Balkans and Russia.

Not only do Western, capitalist democracies abdicate the defence of democracy, they also join communism to depredate humanity. A stark example of this convergence is 'Chinese-style socialism'. This model combines the gulag with the most brutal capitalist exploitation of the workers. Still, the model is tolerated and even praised by Western governments and investors, who are greedy for the 300 million consumers that the Chinese market can provide.

Present-day economic globalisation is promoting a primary type of capitalism, more interested in selling pop music, Coca-Cola or McDonald's than in promoting democracy and genuinely spreading democratic institutions, civil society and human rights. In other words, it is abandoning precisely the values that gave capitalism the strength to defeat communism. In Latin America, in Russia, in Asia or Africa, markets have been liberalised and economies have in large part has been privatised. However, this alone, without democratic institutions, civil society and social solidarity will never create genuine modern capitalist democracies.

The only things that are really being created are more imports to which the bulk of the population has no access, new social inequalities, new monopolies and many klepto-capitalists.

The political rights of individuals, democratic institutions, and social rights – the essence of Western civilisation – are at present being subordinated to the liberalisation and deregulation of the market. In its most predatory version, capitalism's triumph is an obstacle to the rise of a new planetary ethic and of a global society with less social exclusion and more environmental protection. The economy is being globalised, but ethics is not. Today the fashion is the quick profit, instantaneous material gratification and the obsession to participate in the material consumption banquet, at any price and no matter how. These are the components of a Darwinian, competitive, predatory behaviour. All science, technology and economic theory are at the service of this frenzy for quick profits and material gratification that is devouring social rights, as well as the environment.

Communism's fall has allowed the most extreme version of the material progress ideology to be spread to all corners of the earth. This version is a far cry from the successful capitalist models of reconstruction and co-operation that were applied by Britain, France, Japan and Germany during the Cold War. The current economic and financial globalisation model is a predatory process, making the rich dangerously richer and the poor poorer. It also dismantles the social progress that capitalism had attained during the twentieth century, subjects the development of democracy to economic interests, destroys national capitalism and creates precious few jobs.

Economic and social depredation is commonly rationalised as a process of natural selection that throws us into competition and into a fight for our own survival and that of our business in the market, heedless to ethical and social and environmental parameters. The humanist Sir Julian Huxley who defended Darwin's theses at Oxford, used to say that we cannot learn moral lessons from nature, because nature is totally 'amoral'. In one of his debates, he stated: if we take nature before the court of ethics, it will be condemned, since it is morally indifferent. Therefore, the progress and peace of human society depend precisely on not imitating, in human relations, the Darwinian laws of natural selection. As an antidote, Huxley proposed that instead of violent imposition and competitive destruction, we should exercise the principles of self-control and benefit from the help of our fellow men. For this humanist, our ethical concern should be directed not towards the survival of the fittest, but towards making fit the greatest number possible, so that they may survive.[3]

Today, man as an economic and environmental predator is irresponsibly gaining legitimacy. The dominant economic ideology considers that the sum of this depredation is the source of society's material prosperity. This line of argument is the weak point in the ethical, political and environmental veneer of legitimacy that masks this highly predatory new version of post-Cold War capitalism. It forgets that the genetic base of the human species obliges man to be an ethical and social animal or to suffer the revenge of his fellow men who have been the object of depredation.

Promoting the instinctive predatory tendency of man is a dangerous philosophical and political adventure that will lead, in the end, to more violence. The social exclusion of human beings is not the same thing as the natural depredation occurring in a food chain, where one animal species preys on another species, as is the case with lions and zebras. The human groups that are preyed upon are not from a different species, and do not always behave like zebras. A human group that feels permanently excluded from the bare necessities of existence, preyed on by the market and by modernity, will finally turn upon its predators, resorting to varied forms of treatment ranging from delinquency to terrorist fanaticism.

The law of the jungle cannot continue to be applied among

men without serious consequences for the whole community. Our genes, which are programmed for survival in society, begin to protest that we are being excluded, that our family or social group is in danger. Then, when we reach that stage, we too dehumanise our predators, resorting to irrational and violent actions against them, as though they were a different species. When certain limits of human depredation are passed, everyone is the loser, including the predators, and the result is massive social and political turmoil.

The present globalisation model's predatory logic, which justifies financial speculation above job-creating investment, which dismantles social achievements and which, in addition, uses labour-saving technologies, is dangerously increasing social exclusion in many underdeveloped countries that are suffering, as well, an explosive growth in their urban population. In recent years, television has shown innumerable scenes of national depredation, civil wars, drug-financed guerrillas, terrorism and extreme delinquency. We are even witnessing the way in which quasi nation-states of the underdeveloped world, instead of developing, have been imploding into violence as ungovernable entities, converting themselves into a kind of aborted Leviathan, engulfed in infernal struggles, and infected by an emotional plague of pseudo-speciation, in which rival groups deny each other's humanity.

Ungovernable chaotic entities (UCEs)

During the Cold War, it was considered unquestionable that the civil conflicts and the domestic violence of the underdeveloped countries were stimulated and even created by the imperialist policies of the two rival superpowers. Any domestic disorder was, thus, suspected of ideological impregnation. The regions of Asia, Africa and Latin America were seen as gameboards where the struggle for world power between capitalism and communism was being played out.

Today the Cold War has stopped, but the armed battles and acts of terrorism in the underdeveloped world have not only continued, but have multiplied. A wave of armed conflicts has spread over all the continents. Since the Berlin Wall fell in 1989, some twenty-three internal conflicts have emerged (or re-emerged), involving over 50 armed groups. Such violent factions are active in Algeria, Senegal, Angola, Burundi, Congo-Brazza-ville, Liberia, Guinea-Bissau, Rwanda, Democratic Republic of Congo (the former Zaire), Sierra Leone, Somalia, Sudan, Lebanon, Turkey, Colombia, Mexico, Peru, Afghanistan, India, Sri Lanka, Burma, Cambodia, the Philippines, Indonesia, East Timor, Bougainville (Papua New Guinea), the former Yugoslavia, the Caucasus, Tadjikistan, and many other countries.

None of these domestic armed conflicts is related to any world ideological struggle. On the contrary, many of the new barbarian warriors of the underdeveloped world are predatory creatures set off by the demographic explosion and by unemployment, plagued with social, ethnic, religious and cultural resentments. All of the above have been exacerbated by the worsening non-viability of their countries' economies in the face of a new global economy. Social exclusion brings out these social, ethnic, religious or cultural resentments, causing battles to erupt that destroy what little there was of state or nation. This violence has produced 100,000 refugees in Latin America, 7.5 million in Africa, 6 million in Asia and around 4 million in Europe. By the end of the twentieth century, more than 17 million men, women and children had been victims of the outbreak of this worldwide predatory plague.[4] By contrast, in the developed nation-states such as Switzerland, Belgium, Canada, Spain, the United Kingdom or the United States, cultural, ethnic and religious differences do not tear society apart, because material gratification helps to maintain these states' cohesion.

The domestic armed conflicts in the quasi nation-states are veritable *conflicts of national self-depredation*, wherein all respect for the most elementary principles of humanity is lost. There, civil war is combined with massive criminality. Such predatory struggles neither liberate nor dignify any people. They only cause massive physical suffering, emotional damage and genocide.

Some countries that have suffered conflicts of national self-depredation have needed humanitarian interventions from the United Nations or from regional organisations in order to reconstruct civilised life or to relieve human disasters that created thousands of refugees and displaced persons. Such were the cases, for example, of Angola, Mozambique, Ethiopia, Eritrea, Somalia, Rwanda, Burundi, Sierra Leone, Liberia, Afghanistan, Cambodia, El Salvador, Guatemala, Haiti, Congo, Bosnia, Albania, Kosovo and East Timor.

Other countries that have suffered, or still suffer, armed conflicts and terrorism, for example Algeria, Colombia, Egypt, India, Mexico, Pakistan, Peru, the Philippines, Sri Lanka, Sudan, and Tajikistan, have managed to control the situation without international intervention. However, there is no guarantee that their national self-depredation will totally disappear. In many of these countries, violence has become a characteristic of national life, in which the growing criminality, drug trafficking and terrorist armed violence combine into a sinister reality.

In countries where violence has erupted, such as Angola, Algeria, Burundi, Cambodia, Colombia, Congo, Egypt, El Salvador, Guatemala, Haiti, India, Liberia, Mozambique, Nicaragua, Pakistan, Peru, Rwanda, Sierra Leone, Somalia and Sri Lanka, the symptoms of economic non-viability worsened during the 1970s and 1980s. In their economies, based on slightly processed, unprofitably priced primary exports, the populations grew at explosive rates. Food production and the per capita consumption of energy and water lagged far behind population growth. In this manner, the imbalance between population and the physical resources that are vital for social cohesion grew apace. Food insecu-

rity increased. These countries increased their food imports and
became dependent on food aid. At the same time, the lack of
energy security became critical. Some countries lost their self-suffi-
ciency in petroleum, while others raised their imports of that
strategic fuel. All of this coincided with considerable reductions in
the prices of their primary exports, making the real income growth
per capita of these countries equal to zero, as occurred in the
1960s. The incomes of a high percentage of their populations
dropped, and another large segment continued to be born into
poverty.

All these viruses of non-viability caused prolonged periods of
impoverishment, which preceded the violence. For example,
during the twenty-five years prior to its civil war, El Salvador
registered, on average, zero growth of per capita income, while
the population increased at 2.5 per cent per year. Haiti and
Somalia, for thirty years, both registered an average decrease of 1
per cent in the per capita income, with an annual population
growth of 2 per cent and 3 per cent, respectively. In the course of
the twenty-four years preceding the great increase in the terrorist
violence of Shining Path and the Tupac Amaru Revolutionary
Movement (MRTA), Peru had an average yearly income per
capita growth rate of 0.1 per cent, with a yearly population
increase of more than 2.3 per cent. Nicaragua registered an
average yearly decrease of −1.3 per cent in per capita incomes
during the twenty years prior to the outbreak of civil war, while
the yearly population growth was 3 per cent.[5]

In all the countries that suffer today some degree of armed
violence, the average per capita income for the thirty-five years
from 1960 to 1995 grew by less than 3 per cent, which is the
minimum growth needed in order to escape from poverty. In
Algeria, over three decades, the average growth rate of per capita
income was 0.5 per cent; in Angola 0.2 per cent; in the
Democratic Republic of Congo 0.2 per cent; in Sierra Leone 1 per

cent; in Sudan 0.1 per cent; in Mexico 1.8 per cent; in Colombia 2 per cent; in Peru 0.2 per cent. The same tiny income growth rates happened in Liberia, Rwanda, Burundi, India, the Philippines, Sri Lanka and many other countries. The violence was due not only to these low income per capita growth rates, but to the combination of these with an explosive population growth, which exceeded 2.5 per cent per year, and with a deficient distribution of the national income.

The armed struggles caused by national self-depredation can settle into situations of intermittent violence, with repeated armed truces followed by new outbreaks of fighting, in which the warlords divide up or share the monopoly of violence that was formerly the exclusive province of the state. When this occurs, the country has become an ungovernable chaotic entity (UCE).

The UCE is characterised by a collapse of state control over the territory and the population. It is a violent entity where public order no longer prevails either in the cities or in the rural areas. The entire country rebels against central power. Regions, provinces, cities, all lack a representative government and are controlled alternately by military chiefs, warlords, drug traffickers, even thieves, or by an assortment of these. The political process evaporates, legality disappears, representative institutions are replaced by armed forces, or armed rebel groups, or drug-trafficking mafias. The civilian population become citizens of the International Red Cross, Caritas, Médecins Sans Frontières (MSF), hundreds of NGOs, and the intensive humanitarian care of the United Nations.

At this moment, the ranks of the UCEs include Afghanistan, Albania, Bosnia, Burundi, Cambodia, Colombia, Congo (former Zaire), Haiti, Liberia, Rwanda, Sierra Leone, Somalia and Tajikistan. The proliferation of UCEs since 1990 is global reality's most definitive answer to the myth of development. At this moment in history, countries not only do not become newly industrialised, they disintegrate instead.

If the current situation in much of Africa, Latin America and Asia remains unchanged and if the urban population explosion continues, the moment will come in many more states when the inability to satisfy the vital necessities will shred even more their fragile social fabric and erode the relations between the society and the public authority. This will result in greater social, ethnic and religious tension, and will foster the outbreak of new forms of national self-depredation or its resurgence, and the appearance of more UCEs.

Peace enforcement and intensive care

The international community lacks effective mechanisms for dealing with the massive violations of human rights, crimes against humanity and waves of refugees that are caused by the conflicts of self-depredation of the poor countries and their implosions into UCEs. The ineffectiveness of the United Nations Security Council and of its blue-helmeted troops has been accepted as an inescapable fate. The United Nations was designed in 1945 to deal with international conflicts between states. It was not given the means to pacify the contemporary domestic anarchy that began to proliferate in the world.

The ineffectiveness of the organisation is a consequence of the lukewarm political commitment of the great powers, sending out peace missions that did not achieve peace. In civil conflicts, the blue helmets did not intervene to impose a cease-fire and disarm rival bands. They only attempted to guarantee that international humanitarian aid would reach the civilian populations that were victims of the conflict. Paradoxically, they delivered the aid, but did nothing to protect the victims. At the same time as they were distributing food, medicine and shelter, they were witnessing killings, without doing anything to stop them. The fiascos of Somalia, Bosnia and

Rwanda are the clearest examples of this cynical combination of humanitarian aid and genocide.

The lack of political will, on the part of the great powers, to become militarily involved in the protection of the human rights that they are so wont to preach has its roots in the consumerist nature of their own societies based on material gratification. Their citizens are unwilling to sacrifice resources and lives for causes that are not related to their immediate prosperity. Now that Cold War tensions have abated, no one in the Western developed countries wants to see their soldiers die in 'barbarian lands' in wars that they do not understand. The democratic governments of the great powers have an obsessive fear of sending armed contingents, of suffering casualties and of the consequent electoral reprisals. When considering an intervention, most of their military high commands first calculate the possibility of doing so with zero, or nearly zero, casualties. If the casualty estimate is higher than that, they simply do not take part. That is the main reason for their refusal to constitute a permanent peacekeeping force under the United Nations. Every time domestic mayhem breaks out, the debate about intervening or not is reopened in the Security Council. In most instances, the Security Council abandons the country to its fate, granting only humanitarian aid to ease its collective conscience.[6]

One of the few possibilities of establishing a permanent United Nations force for peacekeeping operations, while dispelling the obsessive fear of the great powers that their armed forces might suffer casualties, would be to recruit a mercenary force financed by the Group of Seven and members of the United Nations, placing it at the service of the organisation, as a sort of blue-helmeted Foreign Legion.

The legion at the United Nations' service would be a corps of professional soldiers, recruited and contracted in all the countries of the world. They would have the most modern military technology and combat techniques. They would also count on the powerful logistic, naval and air support of the great powers. Their high command would be subject to the authority of the Security Council and they would be based at strategic points of the globe, with rapid deployment units to defuse conflicts, extinguish domestic fighting, and take as prisoners those who commit crimes against humanity, in order to send them to an international court, such as that recently established in Rome.

The idea of serving in the mercenary forces of the United Nations must be separated from the memory of racist mercenary groups that have fought in recent times against the independence of African countries. Rather, this mercenary force must be associated with the need to have a professional military force at the service of just international causes, such as defending human rights and impeding crimes against humanity. A review of world military history would reveal that mercenaries have been used by many countries to defend causes that they considered just. Professional Swiss regiments have served the kings of France, Napoleon and the Pope, with honour and efficiency.[7] Britain and France have entire regiments of foreign mercenaries, such as the Foreign Legion and the Gurkha regiments. The Latin American wars of independence were also fought with some participation of British sailors and mercenary regiments.

Today, countries with great military traditions, like Britain, the United States and France, have totally professional armed forces. They could be said to be mercenaries with a single national origin, who offer their services for a salary and for a military career. Furthermore, the situation is changing so radically that now it is the African governments that are hiring mercenaries to establish order in their chaotic and violent societies. In South Africa, a transnational corporation called Executive Outcomes is providing security with professional armed brigades to many African states. It is paid in natural resources or in hard currency. Its members are not adventurers, but very disciplined and efficient professionals, who are very careful not to violate human rights.

In the United States, an organisation called Military Professional Resources Inc. (MPRI) has been founded, staffed by more than 2,000 retired US generals and military specialists, whose services may be requested by any country in the world. The largest mercenary military enterprise in the world, specialised in military consultancy and training, this company advised and trained the Croatian and Bosnian armies against the Serbs. It would not be surprising if the United States, in the course of this new century, were to resort to private companies and mercenary units, in order to intervene militarily while avoiding national casualties that would upset public opinion.

There is no reason for the United Nations to exclude itself from this growth in the offering of professional military services. The 'blue mercenaries' formula could dispel the great powers' fears of incurring casualties and, in addition, provide a sword with which the UN could pacify domestic infernos in a timely fashion. However, it is most probable that the organisation will not be strengthened with a military branch but will continue to deteriorate, as an international organisation that has not adjusted to the mission of pacifying domestic wars of national self-depredation.

Indeed, in the high circles of the great powers there is every day less political will and enthusiasm for strengthening the United Nations. The leaders and the citizens of the prosperous northern societies have worn out their compassion for so much misfortune and killing in the underdeveloped world. They have become accustomed to seeing poverty, violence and genocide every night on television. The only country that really could effectively strengthen the United Nations, the United States, does not seem to have the slightest intention of doing so. Its strategists appear to agree that it does not need the United Nations to defend its strategic interests around the globe, since it is at present the only superpower. It thinks that it can do it alone, with its own forces, supported by NATO and its allies across the world.

Faced with the absence of a permanent force at the service of international peacekeeping, as well as with the existence of deficient government forces in many poor countries, it is likely that the trans-nationals will begin to offer private military services and, in the future, hire mercenary military units themselves for the protection of their installations and personnel. Thus, the transnationals would recruit private battalions of *universal soldiers*, trained with state-of-the-art technology, who would provide military services to countries and protect the transnationals' own interests around the world.

Because of the bankruptcy of government power in many poor countries, this trend to privatise military force has already taken hold. Many African governments have contracted transnational services for public safety and for advice on internal security. In Colombia, British Petroleum has privately hired an entire battalion of the Colombian army to protect its installations against FARC guerrillas.

Given that it is practically certain that the United Nations will not be strengthened and provided with the military capacity to quell domestic conflicts, the only way to compensate for this shortcoming would be to establish a preventive system of intensive economic care for poor countries showing symptoms of non-viability in their economies and threatened with the disintegration of their social fabric and with the outbreak of a war of national self-depredation.

This intensive economic care should aim to cancel a large part of the threatened country's foreign debt, offer new credits, massively increase family planning assistance, increase food and energy production and the water supply, and implement measures to avoid social disintegration. Equally important, this intensive care should also design a strategy for modernising the national enterprises, providing greater technological input for production and exports.

These measures would aim to help the economy to become economically viable. Such modernisation cannot be achieved through twentieth-century international co-operation methods, including international aid and the adjustment programmes of the IMF and World Bank. The most efficient manner of solving the modernisation problem of many non-viable economies would be to grant participation in such an enterprise to the transnational corporations and their vastly experienced managers.

Of course, the transnational corporations' participation in modernising these economies would have to be approached as a business proposition rather than as an international aid operation. Otherwise, these enterprises would not agree to take part. The most difficult aspect would be to find formulas that would attract their collaboration. They could, for example, participate through a consulting firm to identify which companies with potential global competitive advantages could be encouraged in a given country. Another formula could assign to a transnational consortium the task of developing a programme for modernising one given industry, in exchange for a part of the returns when the enterprise began to turn a profit. Still another possibility would be to give such a consortium a percentage for the efficient management of one services sector. It might also be agreed to pay the consortium a percentage of the new exports that are created under its supervision. There are many formulas that might be tried, while always keeping in mind the requirement that the economy should begin to modernise and that income levels and the employment rate should show a substantial improvement.

The transnational corporations have not yet envisaged this matter as a future business possibility, and perhaps they will take some time to realise its potential, since they are still enjoying the best possible circumstances, with a worldwide free market devoid of any global responsibility. The majority of transnational corporations see very clearly that world power today has an economic and a technological dimension that favours their businesses without obliging them to bear any international economic, social, or environmental responsibilities. Moreover, in view of the present process of global liberalisation, many transnationals believe that they will have sufficient markets in the future. In China and India alone there are nearly 500 million potential consumers, that is to say, a market almost like those of the United States and Europe combined. A scarcity of global markets in the short run thus does not seem to worry the transnationals.

Nevertheless, if social exclusion and unemployment continue to beset the world, by the second decade of the twenty-first century, only 2 billion of the nearly 8 billion inhabitants of the planet will have high enough income levels to be customers of the transnational corporations.[8] Their clientele will then be only a small global class made up of a major sector of the population of the northern hemisphere, in addition to the small social groups that have high incomes in the poor countries. Perhaps then, faced with this market limitation, capitalism's ability to adapt will lead many transnational executives to realise that the quest for profits is not incompatible with participation through business in international intensive care projects aimed at modernising the non-viable economies and creating global clients.

For the time being, the transnational enterprises are only interested in the clients with the most money and do not seem concerned about the vast majorities with low incomes. Today nobody knows with any certainty whether or not a long-term relationship exists between the low incomes of the vast majority of the world population and the expansion of the market for the transnational corporations. It may become necessary to investigate this matter, because it might help to know whether some day transnational corporations will take an interest in world poverty or not.

Most transnational executive officers, like the bulk of the great powers' political leaders, as well as many technocrats of the IMF and the World Bank, do not feel threatened by the poverty, socio-political turbulence and environmental disasters that are occurring around the world. Rather, they are totally convinced, even though they do not admit it openly, that social and environmental depredation are necessary evils, the price that has to be paid. They believe that the invisible hand of the global market and the new technologies will begin to produce increasing prosperity and solve, in that way, all the problems in the next decades. Despite all the present difficulties, they are firmly persuaded that worldwide prosperity will be achieved and that it will be the result of totally free global competition among all the vested interests.

Given such a predatory world vision, it is currently impossible to establish any system of preventive intensive care to provide timely assistance to non-viable economies threatened with collapse into violence. The main obstacle today is not a lack of financial or technological capacity. It is a lack of an ethic that considers humanity as one single unit. The economic interests may be global, but the ethic is parochial. The reason is that the citizens, the politicians of the great powers and the stockholders of the transnational corporations do not feel the effects of the economic non-viability being suffered by the majority of the world's population. In this respect, the world is not so global. There is still no inextricable connection between the misfortunes of the South and the well-being of the North. The only tie existing today is compassion and the ethical duty of a handful of persons, but this cannot create a global ethical conscience in a world under the influence of a predatory logic.

Within this international context, the great powers will continue to consider case by case the national self-depredation conflicts and the implosion of countries into UCEs. Some countries will be abandoned, like Rwanda and Somalia, Sierra Leone, while others will be supervised but not modernised, such as Cambodia and Haiti. Preventive intensive care will be reserved for those countries that some theoreticians have called 'pivotal states' – that is, countries whose stability must be maintained because their collapse into economic insolvency or anarchy would represent a threat to the strategic interests and the global business of the great powers.[9]

Pivotal states could be Mexico, Brazil, Argentina, Venezuela, Algeria, Morocco, Egypt, South Africa, Kuwait, Saudi Arabia, Pakistan, India, South Korea, Indonesia, the Philippines and Thailand. In these countries, any major socio-political disturbance would doubtless affect the stability of the region, the interests of the great powers, and the interests of the international investors. This 'pivotal states' approach has already been applied with the billion-dollar rescue operations for Mexico, Thailand, Indonesia and South Korea.

CHAPTER 6

SURVIVAL

The decisive factors

Faced with the Darwinian effects of technology and the global market, many countries, as we have seen, are trapped in technological backwardness, primary exports and urban population explosion. Their only option must be simultaneously to lower their birth rate and modernise their production with more technological input. Of course, they have been trying to do this – achieve economic and social development – for half a century, but without success. But even if the modernisation of production and exports proves to be possible, it is a most complex matter that will take a long time to achieve, probably two decades, and in the meantime the nation must survive, avoiding socio-political schisms. For this, it is urgent to establish a balance between population growth and vital resources like food, energy, and water. It is also indispensable to maintain a calm socio-political atmosphere in order to carry out the modernisation process.

If the urban populations of Bolivia, India, Morocco or Nigeria continue to grow, outstripping the production of food, energy, and water, the result will be more poverty, caused by a tremendous demographic pressure on these three crucial resources. The urban demographic explosion will encroach on agricultural lands, thereby making the lack of food security more acute. In the cities, with water scarce or polluted, epidemics will become endemic. Searching for fuel, the population will chop down the forests, causing soil erosion and an even greater decline in food production. With an urban population explosion and shortages of food, energy and water security, there can be no hope of achieving development. Without water, the nation will not have food; no school will be any use if the children are undernourished; no factory will have high production without energy and water; no family can lead a healthy life without sufficient food, water and energy. There will be constant poverty, sickness, unemployment, and rising delinquency, and the social fabric of many poor countries will unravel even further.

Since the dawn of humanity, the minimum physical and social balance required for the existence of a civilised life and social cohesion is that the population does not exceed the supply of vital resources such as food, water and a source of energy. All civilisations have depended on an adequate supply of food, water and energy. The ancient Greeks considered that the basic elements for life were land that produced food, water that purified and gave health, and fire that provided power for human activities. In their view, without food, water and an energy source, there was no viability, either personal or collective, in their *polis*. This fundamental truth still holds in our time, even though the technocratic vision which is nourished by the myth of development has concealed that fact. Now, many poor countries have entered the danger zone as far as their viability is concerned, because these essential elements for life have become scarce and expensive, in the face of urban population growth.

A country can adjust its economy, balance its public expenditures, lower inflation, liberalise and privatise, as Ghana, Morocco, Pakistan, Peru or Zimbabwe have done. Still, if it does not diminish its population growth, or produce adequate food, energy and water for its people, procuring, as well, a strategic international advantage so as to receive more foreign aid and investment as a nation-state, it will always be subject to socio-political instability, and even to violence.

Among a series of recent multidisciplinary studies on population and world resources, the one prepared by the North

American organisation called Carrying Capacity Network (in two volumes totalling 2,600 pages) is noteworthy. These studies conclude that vital resources such as food, energy and water are becoming scarce and expensive in view of the surge in the poor urban population of many countries. Another interesting study, by Goldstone, on revolutions and rebellions in the modern world, explains with great erudition how the British revolution of 1640, the French revolution of 1789, the Central European revolutions of 1848 and the rebellions in the Ottoman empire and in imperial China arose from a lack of political capacity to meet the serious problems that were caused by the constant growth of the population and the shrinking supply of resources.[1]

According to Goldstone, the political instability that is born of the growing discrepancy between population growth and resource levels is analogous to the instability of the tectonic plates of the earth. That is to say, it is known that some day the interaction of these tectonic plates is going to produce an earthquake, but nobody can predict exactly when it will occur. It is not possible, either, to predict when a country will have passed the limit of tolerance of the gap between its population and its resources and when it will implode violently. Even less can anyone predict whether this violence will be disguised as an ideology, a religion, or an ethnicity, or if it will simply be a mixture of anarchy and common delinquency.

Professor Homer-Dixon of the University of Toronto, who has made possibly the most profound studies on resource scarcity and resource security, contends that up until now, the whole focus on problems of the underdeveloped world assumes that the future of the poor countries is determined by socio-economic causes more than by natural causes. According to this researcher, this traditional view is the product of an ideology of progress that since the Industrial Revolution has forgotten nature. He feels that nature is coming back to take its revenge, accompanied by resource scarcity

and climatic change, at the very moment when an urban population explosion is assailing the planet. Homer-Dixon considers that,
in order to have a clearer idea of what will happen socio-politically
in the world, a physical and social theory is needed. This means a
vision in which the destruction of the environment and the
dwindling supply of physical resources for life, in relation to the
growth in population, count for more than do the traditional
socio-economic theories.

By the year 2020, the population of the poor countries will
have nearly doubled, reaching 6.6 billion, and it will be largely
urban. Unless a drastic, unprecedented fall in the birth rate occurs
and is coupled with an unprecedented rise in the supply of food,
energy, water and jobs, the greater part of the population of the
underdeveloped world will live in chaotic cities and megalopolises
with millions of poor and unemployed, without adequate incomes,
beset by malnutrition, pollution and violence.

The fact that poverty in the underdeveloped world is beginning
to change its rural environment to become increasingly urban is a
matter for grave concern, since this new poverty will be more
destabilising than the traditional rural sort. This urban poverty is
concentrated in a space where the lack of food, water and energy is
more acutely felt and where, besides, poverty lives in close
proximity to wealth. It is hardly surprising that today nearly all the
cities of the underdeveloped world suffer a growing plague of
delinquency and that they breed fanatical and fundamentalist
movements.

In Africa and Asia, outbursts of new violence are constant and
entire countries are collapsing in famine, civil war and genocide.
In Latin America, Colombia, El Salvador, Guatemala, Haiti,
Nicaragua, Mexico and Peru have already experienced the scourge
of armed violence and terror. At present, the Latin American
region has the highest crime rate in the world and no one can
guarantee its socio-political stability. History has certainly not

ended with the collapse of communism, despite the assertions of the North American essayist Francis Fukuyama.

For nearly all of Africa, for some countries of Central America, the Andean region, the Middle East and Asia, the great challenge for the beginning of the twenty-first century will not be national development. It will be, instead, national survival. In other words, it will be a question of avoiding further disintegration of the social fabric, as well as preventing the collapse of the project of the nation-state. These countries' governments and their incipient civil societies will have to expend an enormous effort to establish a balance between the supply of food, water and energy and the size of the population. In addition they will need somehow to obtain a strategic advantage that permits them a modicum of international negotiating power, so as to attract foreign capital and technology to modernise their production and facilitate their entry into the global economy.

Thus, increasing the supply of food, water, energy, reducing the rate of population growth and obtaining a strategic advantage become the decisive factors for national survival, as the new millennium begins.

Food

Every country that imports food today should remember that the agricultural land per capita on the planet has diminished by 7 per cent since 1979. The soil does not produce as it used to: this is the result of fertiliser saturation, salinity caused by bad irrigation, and desertification from deforestation. In addition, agricultural lands are being devoured by the unstoppable urbanisation of the planet.

Agricultural production has started to decline in nearly every country. According to the World Bank, in nearly eighty poor countries, food production has lost the race with population growth; world grain reserves at present would cover only 45 days, in the case of a grave food crisis. In the new century, the demand for food will increase by about 3 per cent, while food production will grow by only 2.8 per cent. Today the rate of growth of world production of basic foodstuffs – wheat, corn, soyabeans and rice – has slowed down. Since 1984, the rate of growth of world grain and cereal production has fallen below the rate of growth of the world population, shrinking every year by 1 per cent. The World-watch Institute has warned that to date there is no new technology that would restore the accustomed growth in grain and cereal pro-duction to a yearly 3 per cent and bring about another green revo-lution like the one that saved India and other countries from famine in past decades.[2]

The most populous countries on earth, China and India, will begin to import an appreciable amount of food in the new century, in consequence of their urban expansion and environ-mental problems like soil erosion and increased salinity. These imports will further augment demand and raise the world prices of these products, affecting in this way all the other countries that import them. Then, many poor countries will be forced to beg for more international food aid, to avoid serious political upheavals.[3]

Another phenomenon that will affect the supply of food and influence its price is the expected fall in production of the most important world source of proteins; that is, fish. The seas are being depleted, and many species are becoming extinct. Since 1989, the supply of fish per inhabitant has been reduced by 8 per cent and the catches are smaller, from Iceland to Namibia, from Chile to California. The fishing fleets, with 23 million tons afloat, employ-ing 15 million fishermen and operating with high-definition sonar and gigantic dragnets, are practically clearing the seas of fish.[4]

All these trends are matters of grave concern: by the year 2015, agricultural production would need to have increased by 75 per cent, in order to fill the nearly 8 billion hungry mouths on earth. This does not mean that we have embarked on a course that will carry us into a huge worldwide famine. The specialists hotly dispute that probability. However, they do agree that the world has entered a new era in which satisfying the food demands of the nearly 70 million human beings that are born each year in the poor countries will be more difficult and more expensive than it was in the past.

The experts of the Food and Agriculture Organisation (FAO) of the UN consider that the food security of a country consists in always providing for the population a sure supply of sufficient food for an active and healthy life: sufficient, that is, that malnutrition (under 2,400 calories daily) does not occur. At the end of the twentieth century, nearly 800 million human beings in sub-Saharan Africa, South Asia, Central America and the Andean region do not have food security and depend on international food aid. The countries most afflicted by food insecurity are: Angola, Afghanistan, Bangladesh, Bolivia, Cameroon, Ethiopia, Guatemala, Haiti, Honduras, Liberia, Mongolia, Mozambique, Nicaragua, Nigeria, North Korea, Peru, Rwanda, Sierra Leone, Somalia, Sudan, and Zambia.[5]

Facing future food price increases, the industrialised countries and the NICs, like South Korea, Hong Kong and Singapore, which have low population growth rates and reap large profits on the global market with their growing exports of high-technology manufactured goods, will have no problem in importing food even at higher prices. Neither will there be a problem for the countries that are major exporters of petroleum, like Saudi Arabia, Kuwait or Libya, even if they have high demographic growth, since their large food imports will be financed by their enormous oil revenues. The countries that will have serious problems will be those mentioned above which have the lowest calorie consumption per capita in the world and which can only rely on the unstable income from their primary or semi-processed export products.

Those countries will not be able to import increasing amounts of food. This will make them even more dependent on foreign food aid, thus becoming indigent countries, almost bereft of sovereignty, at the mercy of the donor countries.

For such countries, attaining food security will not be easy, since it does not depend exclusively on their national policies. In fact, the world trend in agricultural trade is to expose national farmers to global competition. The globalisation and the liberalisation of agricultural trade have granted a great power for penetrating the market to the transnational enterprises that produce food. Their very competitive prices eliminate from the competition the farmers from poor countries and erode their national food security policies. This transnational power, moreover, is blessed by the new liberalising rules for agricultural trade promoted by the WTO. These tend to penalise any state intervention to help local farmers and assure the supply of food. To replace food security policies, then, foreign programmes for food aid are promoted, making the poor countries even less able to produce their own food.

In consequence of the world urban explosion and the decline in irrigated land, foodstuffs are becoming veritable strategic resources for foreign policy. They will become scarcer and more expensive and will be used as levers to foster the national interests of the exporting countries. Countries without food security can hardly be sovereign viable states, because they will be at the mercy of external pressures from the countries that supply them with food aid. Thus, little by little, they will turn into beggar states. The only way to avoid this would be to undertake an enormous national campaign to increase food production and diminish food imports and aid, to the greatest possible extent. In this sense, the minimum national goal for countries that today no longer have food security, like Bangladesh, Ethiopia, Guatemala, Haiti, Peru, Somalia, and some others, would be to cease to be net importers of food.

Water

Water to produce food and supply industrial and human consumption is scarce and difficult to access in vast areas of the globe. Some 97 per cent of all the earth's water is saline, only 3 per cent is fresh and three quarters of that is concentrated in inaccessible geographical areas, such as polar regions and glaciers. As a consequence, only a small fraction of the planet's water is both fresh and accessible in rivers, lakes and underground water tables. According to international hydrological studies carried out by the United Nations and the Stockholm Institute for the Environment, even this small fraction is diminishing and in the year 2025, two thirds of the world's population will be affected by water shortage. This is due to the diminishing of the earth's hydrological cycle, caused by the urban population explosion. The symptoms of a water crisis are already visible: underground water, lakes and rivers are shrinking all over the world.

Water security, always a condition for the existence of a civilisation and a nation-state, has begun to disappear in many countries. According to the UN report cited above, already today more than 2 billion people suffer water shortages in more than 40 countries. The World Bank estimates that 1 billion persons already live without enough drinking water and 1.7 billion lack sanitation. The lack of drinking water and sanitation is condemning millions of the inhabitants of underdeveloped cities to sickness and premature death. Having a safe water supply is becoming a decisive factor for national survival.

Even where water is available, the crucial problem is whether it is drinkable. Nowadays, most of the rivers, lakes and streams that flow through large and expanding human settlements are polluted with agricultural pesticides, industrial waste and human excrement. The process of depletion and pollution of the available water is far more acute in the poor countries. Today, half the population of the underdeveloped countries live with a water shortage and suffer diseases related to water pollution, diseases which cause some 25,000 deaths daily. Only 2 per cent of human excrement and industrial waste are subjected to some sort of treatment. The rest, nearly two million tons daily, is dumped in ways that pollute rivers, lakes, seas and underground water tables.[6]

As a result of the explosive growth of cities in the poor countries, farmers are under pressure to produce more food using less water. Poor countries that suffer from a lack of water security will have great difficulty in attaining that goal if the water from their rivers, lakes and underground strata, instead of being channelled to irrigate crops, is directed in increasing amounts to sanitation pipes, industrial use, swimming pools, parks, golf courses and stadiums.

The United Nations considers that the minimum requirement for a healthy and active life is 2,000 cubic metres of drinking water per year.[7] The countries with water resources already close to that limit for survival are: Algeria, Burundi, China, Egypt, Ethiopia, Haiti, India, Jordan, Kenya, Morocco, Oman, Pakistan, Peru, Rwanda, Sri Lanka, Yemen and Zimbabwe. In the year 2005, many cities of these countries will have only half the amount of water that they had in 1975. The most seriously affected will be the large cities, such as Algiers, Amman, Cairo, Casablanca, Lima or Tunis.[8] The water shortage in these underdeveloped countries is a problem not only of physical quantity, but also of personal income and of the equal adjustment of consumption to the national environmental reality. Cities where the greatest part of the population is concentrated in huge arid zones cannot continue to grow in the shady green style of Beverly Hills.

The countries identified as having a chronic water shortage will not only be forced to import the food that, without water, they are unable to produce, but also be exposed to epidemics. Megalopolises built in arid zones, such as Cairo or Lima, will be the first in the new century to suffer the devastating effects of the scarcity of water on the inhabitants' quality of life. Soon they will be joined in their misfortune by other huge cities, because by 2020 at least 70 per cent of the planet's population will be urban and the planet's water cycle will be insufficient. By about 2020, the number of persons living in countries with insufficient water will have reached nearly 3 billion.

There is no doubt that the problem of water supply, both for agricultural use and for the urban population, will become more political in the twenty-first century. This will be due, in part, to the insatiable thirst produced by the world urban demographic explosion and, in part, to the need for irrigation to produce more food for the cities. In an urbanised planet, with nearly 8 billion inhabitants by the year 2020, water will be as strategically vital for living as petroleum. Hence, it would not be at all surprising if its scarcity were to provoke national and international upheavals reminiscent of the oil crises of the twentieth century. It is very possible that capturing water sources or polluting reservoirs may become strategic objectives of war plans and of terrorist attacks.

If countries that have considerable urban expansion and that are suffering droughts, desertification and water shortage do not begin to remedy the situation in the first decades of this century, they will have to confront a growing water crisis. The worst situation will be that of countries that already have a low rate of water consumption per inhabitant and that concentrate the better part of their population in mega-cities located in arid zones, as is the case of Algeria, Egypt, Iraq, Jordan, Kenya, Lebanon, Peru and Syria.

It is also very possible that water shortage may cause domestic upheavals in mega-countries like China and India, where urban expansion is continuing at full steam. In China, there is already a dire scarcity of water in nearly twenty-two large cities. Millions of Chinese are migrating towards the coastal cities, depopulating the countryside of farmers, and consuming in the cities part of the water that was used for food production.[9] India is in no better condition. Drought and lack of water are persistent, because of soil erosion caused by deforestation which is itself caused by the search for firewood. This is also causing serious problems for the agricultural sector and for the mega-cities of India, making it necessary to use increasingly the exhausted volumes of the Brahmaputra and the Ganges rivers.

Disagreements over the use of the great international rivers can generate international conflicts. The waters of the Tigris and Euphrates rivers, which are being dammed by Turkey to irrigate the region of Anatolia, are also vital for the survival of Iraq and Syria. If the three countries do not reach a tripartite agreement on their use, this failure may very well cause a conflict in the future. In the case of Israel, Syria, Jordan and the Palestinians, agreement on the distribution and use of the Jordan river, which have not yet been defined, would doubtless be a fundamental condition for a durable peace among them. A conflictive situation could also arise around the use of the Nile's waters by Egypt and Ethiopia, since the dam that the latter country is planning to build would use part of that river's waters, which have been vital to Egypt since Pharaonic times.

For the countries that presently register the lowest water consumption per capita in the world, that have a high population growth in great arid zones, and that suffer chronic droughts and have no water security, like Egypt, India, Jordan, Kenya, Morocco, Peru and others, water security should be a strategic objective for survival. These countries have no other choice, if they are to preserve their fragile social cohesion, than to adopt immediately policies for water security. They must search for new sources of water, devise measures to save, purify and recycle the available water and also share out this national resource fairly in accord with their environmental reality.

No one should doubt that in the new century a very low quantity and quality of water available per capita will be a clear international indicator as to which countries are definitely non-viable, since they lack the most elemental resource for survival on the planet.

Energy

All the poor countries try to copy the consumption patterns of the wealthy industrialised nations. The urban modernity they try to emulate implies a growing consumption of petroleum, a non-renewable and polluting source of energy.

Due to these consumption patterns and to urban population growth, the voracity for petroleum of the underdeveloped countries is such that countries that once were self-sufficient and even medium exporters of oil have become net importers of it, endangering their energy security. Even countries that have gas deposits and other fuel sources cannot avoid importing oil, because of the difficulties involved in converting all the industry and transport of a country to other sources of power. This is why it is calculated that the oil consumption of the underdeveloped countries will continue to rise by 5 per cent per year, bringing the demand for petroleum in the poor regions of the planet to over 50 per cent of the total world demand this century.[10]

Despite the high rise in demand, there are no indications of a possible cutback in the world oil supply, because the world reserves will suffice until the middle of this century, unless a major international conflict should break out affecting the Gulf reserves. Nevertheless, the prices of petroleum and of energy will rise gradually in the long run, not only because of growing demand, but also because extraction costs will be higher. Additional costs will also arise from the geographical location and the geological condition of the new and the old deposits, as well as from the use of environmentally friendly technologies in processing fuels and in all domestic or industrial transport systems.

It would appear that this price increase will not be sudden, nor will it produce world energy crises like those of 1973 and 1980. Rather, there will be a series of national shortages and energy crises that will affect many of the poor countries. This will be especially hard on those that have rapid urban population growth and that, in spite of having increased their oil imports, continue to register the lowest annual per capita energy consumptions in the world, between 200 and 600 kilos (the petroleum consumption per capita of an Asian NIC is almost 2,000 kilos). According to the United Nations, the following countries are in this situation of fuel shortage: Afghanistan, Bangladesh, Cambodia, Congo (the former Zaire), Côte d'Ivoire, Cuba, El Salvador, Ethiopia, Ghana, Guatemala, Guyana, Haiti, Honduras, India, Kenya, Laos, Morocco, Mozambique, Nepal, Nicaragua, Paraguay, Peru, Senegal, Sri Lanka, Sudan, Tanzania, Tunisia, Zambia and Zimbabwe.[11]

Without doubt, the countries mentioned above may have hydroelectric potential or deposits of uranium or gas. However, the enormous investments required in order to construct dams, hydroelectric and nuclear plants would hardly allow these countries to replace their oil imports in the near future. According to the World Resources Institute, as a result of urban population growth the future energy requirements of the majority of these oil-importing countries, in this century, would demand gigantic investments, as high as a trillion dollars. Obviously this sum greatly exceeds the foreign capital available for investment in new refineries, pipelines, dams, central power stations or alternative sources of energy.[12]

The countries that have the lowest per capita energy consumption in the world also import food, all sorts of consumer goods, and capital. This will make it ever more difficult for them to continue importing increasing amounts of oil, in order to increase the per capita consumption, given the dwindling value of their exports (based on primary products and manufactures with low technological content). Energy shortage could then become a structural crisis and turn many of these countries definitively into non-viable national economies. The rural population will continue to depend on firewood as domestic fuel, aggravating the deforestation process and soil erosion, and deepening the lack of food security. Without adequate energy supplies, these countries cannot provide water, sanitation and transport for their expanding cities. Even less will they be able to attract foreign companies to increase the technological content of their exports and improve their archaic position in the global economy.

The scarcity of petroleum and its rising cost will turn the already notable social inequality into a gaping abyss, as only a small sector of the society will have the necessary personal income to buy enough energy to maintain the consumption patterns which it copies from the rich societies of the North. The majority of the population, on the other hand, who live on less than a dollar a day or with very low incomes, will not be able to keep pace with increased consumption of commercial energy, and their material living conditions will drop even further below standard.

The countries, like El Salvador, India, Kenya, Morocco, Pakistan and Peru, currently with the lowest energy consumption per capita in the world and a growing urban population, must try to turn this energy shortage around. Otherwise, they will condemn their populations to hereditary poverty and become non-viable countries. They urgently need to establish effective national policies for energy security. This means that their populations must be assured access to a large enough quantity of commercial energy to satisfy at least the basic necessities in food, housing, communications and drinking water.

To reach energy security, it is essential to save energy and to maximise the exploration and exploitation of the renewable and non-renewable energy sources in the country. Another prerequisite is to relinquish the dream of imitating the consumer society of the wealthy industrialised countries. In the long run, the national

energy security issue will act as the catalyst that produces another
type of society, one that no one can foresee. It is clear, however,
that it will be contrary to the paradigm of the consumer society,
which neither poor nor rich are willing to abandon yet in any
country in the world.

Although it is understandable that no one in the poor countries
wants to relinquish the dream of the rich countries' consumer
society, it must be admitted that nothing is more irrational than to
try to globalise this ideal, based as it is on the intensive use of a
highly polluting source of energy: fossil fuels. The poor societies
would have to reach a per capita consumption of fossil fuels that
approached the North American or the European rate. Try to
imagine a great part of the African continent, Asia including mega-
countries such as China, India and Indonesia, and all of Latin
America, that is, three-quarters of the world's population, consuming between 4 and 7 tons of petroleum per capita and
expelling into the atmosphere millions of tons of greenhouse gases,
to add to those accumulated since the Industrial Revolution. The
atmosphere would be still more loaded with carbon dioxide,
methane, nitric oxide and chlorofluorocarbons; the planet's temperature would continue to rise. The effects of this would be felt
in still more severe climatic changes than those at present,
producing more droughts and torrential rains that would affect
agricultural production. In addition, the damage to the polar ice
caps would raise sea levels, flood the coastal countries and obliter-
ate many islands.

Today, the paradigm of the consumer society, riding on the
intensive use of such a polluting source of energy, has placed our
civilisation in one of the most serious dilemmas ever experienced
in the pursuit of material progress: a low per capita consumption of
fossil fuels leads to the economic non-viability of the nation, but a
high consumption per capita by all the nations would lead to the
virtual non-viability of civilisation. For the first time, the energy

dilemma unequivocally places environmental issues at the centre of the socio-political destiny, not only of countries, but of the entire human race.

Population stability

While food, water and energy are becoming more strategically important, scarce and expensive, by the year 2005 the most radical historical change will have occurred in the occupation of the planet by the human species. For thousands of years, the planet's population was settled in rural areas. In a short time, that will no longer be the case. About 55 per cent of the planetary population, some 3.35 billion people, will live in thousands of cities, spread over all the continents. The most worrisome fact is that 90 per cent of this explosive city growth will occur in very poor countries lacking in food, energy and water.

Every day the urban population in the poor countries swells by some 150,000 people. In the year 2020, the present population of the great majority of these countries will have nearly doubled, reaching a total of 6.6 billion people. Almost 70 per cent of them will be living in cities. More than five hundred cities with more than 1 million inhabitants each and some forty megalopolises that will have between 7 and 20 million people each will burgeon in the so-called developing world.[13] Kinshasa will reach 7 million, Lima will have 10 million, Manila 14 million, Cairo 14 million, Dhaka 18 million, Delhi 18 million, Karachi 21 million, and Shanghai 22 million; hundreds of secondary cities in those countries will pass the 1 million-inhabitant mark as well.[14]

This urban explosion will require the backing of mega-economies with enormous capacities to produce foodstuffs, energy and water. Without these three resources, cities become veritable human nightmares and environmental time bombs. When London's population reached just 5 million or when New York's arrived at 7 million, behind each of those megalopolises was a powerful industrial development, which absorbed great quantities of labour, coupled with a huge national and international market. They were also backed by prosperous agricultural sectors and cheap and plentiful energy in the form of coal and oil. Lastly, the poor and the unemployed in those cities had the option of emigrating massively in search of their fortune, to the colonies or to the Far West.

In contrast, behind cities like Casablanca, La Paz, Luanda or Maputo and future megalopolises such as Cairo, Dhaka, Karachi, Lima or Manila there is no modern national industrial drive that exports and employs. Even less have any of them an agricultural sector that produces adequate quantities of food, or vast oil reserves. The underdeveloped megalopolis possesses only the inadequate resources from the country's barely processed exports. It depends on foreign food aid, suffers water shortages and has to import increasing amounts of oil and food.

The countries with the greatest urban growth in the world and also with the lowest world consumption of food and energy per capita are: Bangladesh, Bolivia, China, El Salvador, Ethiopia, Ghana, Guatemala, Haiti, Honduras, India, Kenya, Liberia, Morocco, Nicaragua, Pakistan, Peru, Philippines, Rwanda, Sierra Leone and Tunisia. In this group, China, El Salvador, Ethiopia, Ghana, Haiti, India, Kenya, Morocco, Pakistan, Peru, Rwanda, plus Tanzania, Uganda and Zimbabwe, have in addition the lowest per capita consumption of water in the world.[15]

Excepting China and India, all these countries are caught in a trap consisting of technological backwardness, scarcely transformed exports, and low rates of productive foreign investment; their growing urbanisation increases dangerously the imbalance that these countries already have between the population and the vital resources of food, water and energy. In the near future, this imbalance will become still more acute, because the population will double by 2025. If this situation remains unchanged, the above-mentioned countries will remain among the poorest national societies on the planet. Many of them may be driven into economic non-viability.[16]

The underdeveloped countries with urban population explosion and low per capita consumption of food, energy and water need their GNPs to grow steadily by an annual rate of at least 7 per cent during the next ten years, if they are to catch up with the urban population explosion, create employment and attain food, energy and water security as well. However, it is estimated that the GNP of most of these countries, because of their dependence on primary exports and low-technology manufactures, will grow by only 3.5 to 4.5 per cent per year. When population growth is taken into account, this will be reduced to an actual annual growth rate of between 1.5 and 2.8 per cent.

As recently as the 1970s, many strategists held that a large population was the symbol of a powerful nation. Maoist China defended this thesis in the seventies, denouncing birth control as a sinister imperialist plot. Later, reality forced the Chinese leadership to perform an about-face, towards a totalitarian demographic policy of one baby per family and compulsory sterilisation. Brazil too was a defender of the 'large population' thesis. Three decades ago its political leaders claimed that with 200 million inhabitants in the year 2000, it would be assured its credentials as a great power. In the end, overwhelmed by the growth of poverty in its great cities, it embarked on a voluntary family planning policy.

Today nobody dares to equate a large population with national power. Nevertheless, there is still a trend to downplay the high growth of the urban population as a determining factor in poverty. In general, there is a lack of concern about population growth, based on the belief that many underdeveloped countries have fortunately begun what the demographers call 'demographic transition', which means a decline in fertility, making for a more equal relation between births and deaths. A significant reduction in poverty, based on this calculation, is exaggerated, if account is taken that the demographic growth in the major cities of those countries will continue to be 150,000 inhabitants each day, well into the twenty-first century. This is an environmentally and socially explosive rate of population growth.

Research carried out by the United Nations Population Fund has positively shown that a reduction in the size of the family contributes to lowering the infant mortality rate, and to improving education, health, nutrition and living standards in general.[17] It is not pure chance that South Korea, Taiwan, Singapore, Chile, Argentina and Uruguay, which have the fewest poor inhabitants among the so-called developing countries, also have the lowest population growth rates.

Despite this evidence, there are no effective population policies in El Salvador, Kenya, Pakistan, Peru, Sri Lanka, Zimbabwe, Morocco and almost all the other countries that combine the world's highest rates of urban population growth with its lowest per capita consumption of food, energy and water. Commonly in these countries, predictions and discussions about the future behaviour of the economy and society fail to take into account the population factor. It would seem that there is no national awareness of the impossibility of satisfying the growing human needs resulting from an urban population explosion that is over 2.5 per cent per year – given these countries' current dependence on raw materials and low-technology exports. There is no perception either of the fact that the urban demographic explosion will produce a great supply of workers that modern technology cannot easily absorb.

Behind this rather indifferent attitude may lie traditional cultural values, and the belief that resources are infinite and that the relation between environment and population will not affect the future of the nation. Economic traditionalism, coupled with a religious fundamentalism that considers any population policy an attack on life, ultimately blocks actions of the civil society and of the state in favour of voluntary and democratic family planning. Traditionalism and fundamentalism preclude measures to promote gender equality and to allow the less favoured sectors of the population access to the same methods that the most favoured sectors use to reduce their fertility.

Strategic advantage

During the Cold War, the countries of the so-called Third World acquired in different degrees a strategic advantage, thanks to the interest of the superpowers in gaining friends and allies. The non-aligned policy provided some of these countries with a strategic advantage that permitted them to manoeuvre between the two blocs and obtain assistance from both. The clearest example of this was Yugoslavia. That country was virtually a byproduct of the Cold War. Its estrangement from Moscow, coupled with the strategic significance of the Balkan region, earned it ideological tolerance and great financial support from the Western powers, which gave it room to try out an economic model based on self-management. When the Cold War ended, so did the financial subsidy. Yugoslavia became non-viable and imploded in a bloody war.

To a lesser degree than Yugoslavia, India, Indonesia, Algeria, Egypt and even Peru also obtained political support, some economic advantages and equipment for their national defence, by manoeuvring between the two rival blocs. The other option for securing a position of strategic advantage during the Cold War was to throw all one's support behind one of the superpowers. An example of this extreme position was Cuba, which fought for the interests of the USSR. This alliance provided Cuba with a petroleum subsidy and high prices for its sugar exports and permitted it even to play a geopolitical role in Africa.

Today, with the end of the Cold War, most underdeveloped

states have lost the strategic importance that made them recipients of aid or investments. International aid has considerably diminished, and the poor countries do not receive productive foreign investments in any significant amount. Moreover, in political terms, they no longer receive the special, non-reciprocal and differentiated treatment they used to. In consequence, they have to compete on equal terms with the developed countries. By the end of the twentieth century, almost all of the poor quasi nation-states had been strategically abandoned to the mercy of the global market and the technological revolution's process of natural selection.

Within this trend, only a few of the so-called developing countries still have strategic advantages, as mega-exporters of oil or food, two resources that are becoming more important every day because of the world demographic explosion. Among these fortunate few are Saudi Arabia, Kuwait, the Arab Emirates, Dubai, Venezuela, Nigeria, Iran, Iraq, Mexico, Kazakhstan and Azerbaijan, great exporters of petroleum. They also include Argentina, because of its position as one of the world's leading producers of grain, as well as those countries that are situated on straits or canals that are vital to the world economy – Iran, Oman and Saudi Arabia, bordering the Straits of Hormuz, through which passes a high proportion of the world's petroleum, and Egypt and Panama, as a large percentage of the world trade in goods transits through their canals.

Ironically, at the end of the Cold War, the only source of strategic advantage for some countries is the danger their instability represents for their rich neighbours. Indeed, some rich countries have no choice but to help their poor neighbours so that they do not become unstable, thereby avoiding waves of clandestine immigrants or refugees entering their own territories. In addition, the rich states help their poor neighbours to become buffer states, that is, territories that serve to contain the illegal immigration from other, even poorer countries.

Just as the Roman Empire created *limes* (or buffer zones) to contain the barbarian tribes, many rich countries will try in the twenty-first century to contain the new barbarians by stabilising their poor neighbours with credits and investments so as to turn them into buffer states. Being a poor state that borders on a prosperous country or region will gradually grow more remunerative as the economic non-viability of the countries of the South creates increasing numbers of illegal immigrants and refugees.[18]

The countries of the Maghreb are currently acquiring this type of strategic advantage as the poor neighbours of Mediterranean Europe. These countries will double their population by the year 2020, causing a further rise in unemployment – which already affects 40-50 per cent of their young people. Sooner or later, Europe will have to stabilise these countries, not only in order to contain fundamentalism and the illegal emigration from the region, but also to convert them into buffer states for sub-Saharan Africa. The population of the latter is growing explosively and processes of national disintegration have already begun there, causing waves of illegal immigrants and refugees to head towards Europe.

Mexico is also acquiring a strategic advantage as a buffer state. The United States has chosen this country as a partner, in order to stabilise it and to avoid being affected by immigration both from Mexico and from Central and South America. The strategic advantage enjoyed by Mexico as poor neighbour and buffer has been confirmed with the creation of the North American Free Trade Area (NAFTA) and with the colossal financial rescue of Mexico by the United States government to prevent its latest bankruptcy in 1998. Only time will tell whether the United States has made a good investment. Until now poverty has not abated in Mexico, and the illegal northward migration.

With these few exceptions, the deterioration of the strategic situation of what has been called the Third World is such that now the poor countries of Latin America, Africa and Asia have to fight for markets and foreign investments against the former rivals of capitalism, China, Russia and eastern Europe. Those ex-Communist countries are, ironically, more strategic than the underdeveloped countries, because if their transition to capitalism should fail, the rich countries' stability would be threatened with greater unlawful immigration, mafia-type intrigues, trafficking in nuclear material, and even waves of refugees from the ensuing civil wars.

Today, the vast majority of poor countries of Latin America, Africa and Asia do not represent a direct threat, because of their poverty or non-viability, to any powerful neighbour. Nor do they possess any strategic advantage to increase their international negotiating power. Their only option is to use their diplomacy as a strategic instrument to obtain foreign resources to permit them to survive.

By means of effective diplomacy, these countries will have to persuade the industrialised countries to increase their bilateral aid for the purpose of carrying out effective national policies in the areas of family planning, food, energy and water security. In addition, they will need to obtain the cancellation of a good part of their debt burden, and to display an outstanding capacity for negotiation with the IMF, the World Bank and the WTO so as to secure temporary systems of special and differentiated treatment.

In relations with neighbouring countries and countries in the same region, to serve as a real strategic advantage diplomacy should propose initiatives to monitor the acquisition of new systems of arms that might destabilise the region strategically and unleash a costly arms race that would be fatal for these countries' national viability. A climate of neighbourly and regional peace and stability will be a useful base for creating an adequate *national* defence policy. This in itself would be an additional strategic advantage, because it would allow resources to be channelled away from national defence and towards strategic factors for survival, such as food, energy and water security.

Defence spending should be no higher than absolutely necessary: on efficient, but not costly, modern strategic defences. Nothing is more absurd than countries that compete in the world poverty rankings and at the same time are obsessed by an outdated conception of national power. It would appear that the military high commands of these countries are unaware that the most important component of national power in our times is not offensive weaponry but the economic–technological factor. Today, Germany and Japan (no longer military powers) base their strength on their economies and wield more international clout than Britain, France and Russia, which are nuclear powers. Cuba is a heavily armed country that projected its military power as far as Africa. Nevertheless, today Cuba is marginal compared with Taiwan, whose exports penetrate all the world's markets, though it has just recently been bold enough to acquire a few ultramodern supersonic planes.

Besides, the world's new strategic situation shows that since the end of the Cold War, from 1989 to 1997, the armed conflicts between countries have diminished and the number of internal civil conflicts is increasing, as a result of a process of national economic non-viability. Of the twenty-three armed conflicts that broke out between 1989 and 1999, only five were between states (the Gulf War, Armenia–Azerbaijan, Ethiopia–Eritrea, Peru–Ecuador, Serbia–NATO). The remaining twenty-five were internal conflicts. The majority of these conflicts took place in countries with symptoms of economic non-viability. Peru and Serbia were the only countries to suffer both types of conflict.

The high commands of the poor countries' armed forces need to understand that a country that exports raw materials and barely processed products, with high rates of demographic explosion, unemployment and poverty, and that receives food aid to alleviate the population's malnutrition, has no viability to support a war, excepting civil war, which is proof of its own non-viability.

The pact for survival

Within the countries of Latin America, Asia and Africa that combine the world's lowest per capita consumption of food, water and energy with the world's highest rates of urban population growth, a particularly clear example of physical and social imbalance is the case of Peru. During the pre-Columbian era, a great part of the population in what is today Peruvian territory was settled in the Andes, maintaining a delicate balance of food and water. This equilibrium was even conserved by the Incas, who moved the population from valley to valley through a system of forced migration called *mitimaes*. This balance did not deteriorate when the territory became part of the Spanish empire, because the brutality of the conquest and the colonisation significantly reduced the population of the Andes.

However, during the republic, and especially during the last half-century, the Andean population began to grow at high rates and to move massively towards the cities of the Pacific coast. This produced an urban population explosion in an inappropriate natural environment, an arid region with narrow valleys and little water either to supply great cities or to produce food for them.

Lima is growing at an average of more than 100,000 inhabitants a year and other large coastal cities also register very high growth rates. In general, the urban growth in Peru is 2.5 per cent. Lima will reach 10 million inhabitants in the year 2015 and other cities will pass 1 million. Already the coastal cities have spread out over agricultural lands, and taken the water that should be used to produce food. This makes it necessary for Peru to import increasing amounts of food, and to depend on food aid, for lack of land and water. Peru now appears in the world statistics as lacking food and water security, having one of the lowest consumptions of calories and of water of Latin America and the world. The huge coastal urbanisation has caused Peru to lose energy security as well. Self-sufficient in energy in the past, Peru must now import petroleum and is one of the countries with the lowest energy consumption per capita in Latin America.

The dislocation of the Andean population towards a kind of cancerous urban growth, in an arid environment, will be the greatest challenge of all for the viability of Peru as a nation in the new millennium. The age-old relation between environment and population has shown that the arid, mountainous and jungle territory where the project of the Peruvian nation-state is situated is not appropriate to sustain huge cities, least of all in arid zones that resemble nothing so much as a moonscape.

If the whole group of countries that, like Peru, want to achieve socio-political stability so as to modernise their anachronistic position in the global economy, within the next decade they will have to reduce their urban population growth, while at the same time increasing their food production to attain food security. In addition, they will have to augment significantly their water supply, not only for the purposes of food production, but also to provide sanitation and raise health standards in their insalubrious megalopolises. Finally, they need to increase substantially their energy supply, both to satisfy the growing household demand for energy caused by the urban explosion, and to create industries and jobs, as well as to modernise and expand all the urban services. To achieve this, the calorie consumption of the inhabitants of these countries will have to be increased from the barely 2,000 calories per capita at present to almost 3,000. The low fuel energy consumption, which is less than 500 kilos of petroleum per capita now, will have to be doubled; water consumption, under 2,000 cubic litres per capita today, should be at least quadrupled.

As we begin the twenty-first century, the growing physical and social imbalance between food, energy, water and population is ignored in the national agendas of the underdeveloped countries. Indeed, the majority of these countries' governments, and their technocracies, evince an utter lack of concern at the problems. The myth of development is so deeply rooted in the collective subconscious of the political classes that they think that they only have to set in motion the economic and financial policy that is in fashion, and has been dictated by the great economic powers, the transnationals and the international economic and financial organisations. They do not realise that the technological revolution is making anachronistic the only two comparative advantages their countries possess, to wit, abundant unskilled labour and natural resources. Nor do they understand that this process will gradually intensify their condition as non-viable national economies, as quasi nation-states, frustrated national projects. Even less do the politicians and the technocracy of the underdeveloped countries perceive that their countries could become ungovernable chaotic entities, as a result of the process of non-viability of their economies and of a growing physical and social imbalance produced by the urban explosion of their poor societies.

Today, after more than fifty years of applying a variety of development theories and policies, the real per capita income in more than seventy so-called developing countries is lower than it was twenty years ago. Of a population bordering 5 billion in the underdeveloped world, around 3 billion survive on only two or three dollars a day, and another 1.3 billion in extreme poverty can no longer even feed themselves, living on less than one dollar a day. This reality is an invitation to discard the myth of development, abandon the search for El Dorado and replace the elusive agenda of the wealth of nations with an agenda for the survival of nations. It is now crucial to stabilise the urban population growth, and to increase the supply of water, energy and food. The achievement of this physical and social balance is not related to any ideology. Therefore, it should be possible to develop a 'pact for survival' among all the political leaders in any poor country where alarming symptoms are apparent.

One fundamental prerequisite to bring about pacts for survival in the poor countries is that they have genuinely democratic regimes. The pacts for survival should be the outcome of broad national dialogues and of democratic consultations encompassing governments, political parties, business circles, workers, academic communities and civil society in general. It must be admitted by all that dictatorships and other non-democratic forms of government are regimes that will only confirm poor countries in their non-viability; this is because they do not permit the dialogue and the agreement that are indispensable if we are to find adequate answers to the difficult challenges ahead. Non-development is, precisely, a history of authoritarianism, corruption and exploitation that has not allowed the country to go beyond cheap labour and primary exports.

The pacts for survival should not only be *established* through dialogue and democratic consultation, but they should continue to function permanently on that basis. Only through such a constant exercise will it be possible to adjust the pacts for survival to the challenges that the market and the technological revolution will pose along the way for those countries currently infected with the virus of non-viability, technological backwardness, increasingly un-remunerative exports, and growing physical and social imbalances.

NOTES

I The Twilight of the Nation-State

1. See Machiavelli, *The Prince*; Hobbes, *Leviathan*; and Locke, *Two Treatises of Government*.

2. Paul Kennedy, *Preparing for the Twenty-first Century* (Glasgow: Harper Collins, 1991), pp. 37–134.

3. Mostafa Rejai and Cynthia H. Enloe, 'Nation-State and State-Nation', *Perspectives in World Politics* (London: Croom Helm, 1982), pp. 37–46.

4. See Robert H. Jackson, *Quasi States, International Relations and the Third World* (Cambridge University Press, 1990).

5. Kennedy, p. 126.

6. See R. O. Keohane and Joseph S. Nye, Jr., *Transnational Relations and World Politics* (Harvard University Press, 1966), pp. ix–xxix.

7. Susan Strange, *Casino Capitalism* (Basil Blackwell, 1986), pp.1–24.

8. United Nations *Global Outlook 2000* (UN Publications, 1990), p. 139.

9. Ibid., p. 150.

10. Ibid., p. 155.

11. The need for such a system was proposed by the author, as Ambassador Representative of Peru, to the Second Meeting of the Preparatory Commission for the United Nations Conference on Environment and Development, Geneva, 1992.

12. Alan Tonelson, 'Super Power without a Sword', *Foreign Affairs* (Summer 1993, Vol. 72, No. 3).

13. Colin Powell, 'Challenges Ahead', *Foreign Affairs* (Winter 1992–93).

14. *An Agenda for Peace*, report of the Secretary General of the United Nations A/47/272-S/24111, (17 June 1992).

15. Horacio Godoy, 'La Crisis del Estado Nacional Contemporáneo' (The Crisis of the Contemporary National State), in *América Latina 2001* (Bogota: 1976).

16. Eugene Linden, 'Exploding Cities of the Developing World', *Foreign*

Affairs, January–February 1996.

17. Kenichi Ohmae, 'Rise of the Region State', *Foreign Affairs,* Spring 1993.
18. Matthew Horsman and Andrew Marshall, *After the Nation State* (Harper Collins, 1994), p. 94.

2. Global Empowerment and National Impoverishment

1. United Nations, *World Investment Report 1995,* p. 8.
2. Ibid., 'Overview'.
3. Ibid., pp. 3–41.
4. Ibid., p. 8.
5. See the speech and proposal of Peru to the Second Meeting, Preparatory Commission for the United Nations Conference on Environment and Development, Geneva, 1992.
6. Paul Krugman, 'Emerging Market Blues', *Foreign Affairs,* (July–August 1995).
7. Among the critics are professors of Economics at Cambridge Ha-Joom and Robert Rowthorn; Nobel prizewinners for Economics, John Tobin and Maurice Allais; the Professor of Economics at Stanford, Paul Krugman; and the Professor of Economics at London and Manchester, Paul Ormerod. Criticism has also come from investment bankers of Wall Street, Felix Rohatyn, George Soros and Warren Buffet; the *New York Review of Books* (April 1996); Oxfam, the Ecumenical Council of Churches, the Latin American Episcopate, the United Nations Development Programme, and UNICEF.
8. UNDP, *Human Development Report, 1996*; *Human Development Report 1997* (New York, Oxford University Press, 1996, 1997). *World Bank Development Report, 1998.*
9. Paul Ormerod, *The Death of Economics,* (Faber and Faber, 1995).
10. Ibid.
11. World Investment Report, op. cit. pp. 21–3.
12. UNDP, 1997, pp. 146–59.
13. *World Bank Report, 1998.*
14. UNDP, 1996, pp. 24–30; UNDP 1997, pp. 1–17.
15. Ibid., pp. 142–3.

16. Ibid., p. 146.
17. Ibid., p. 147.
18. Ibid., p. 178.
19. United Nations, *Global Outlook 2000* (UN Publications, 1990), pp. 146, 178, 192.
20. Ibid., p. 194.
21. Ibid., p. 198.

3 International Darwinism

1. Paul Ormerod, *The Death of Economics* (London: Faber and Faber, 1995), p. 14.
2. Ibid., p. 12.
3. Peter Passel, 'The Poor get Poorer in the US', *International Herald Tribune* (30–31 March 1996); Ethan B. Kapstein, 'Workers and the World Economy', *Foreign Affairs*, July–August 1996; UNDP, *Human Development Reports 1993–1997*.
4. Richard J. Barnet and John Cavanagh, *Global Dreams* (New York: Simon and Schuster, 1994), p. 382.
5. Ibid., p. 177.
6. Ibid., pp. 369–73.
7. Peter F. Drucker, 'The Changed World Economy', *Foreign Affairs*, Spring 1986.
8. Susan George and Fabrizio Sabelli, *Faith and Credit: The World Bank's Secular Empire* (Penguin Books, 1994), pp. 58–73. Cambridge Professors Ha-Joom, Robert Rowthorn. Nobel Prize winners Tobin, Allais. Investment bankers of Wall Street, George Soros and Warren Buffet.
9. UNCTAD, *Trade and Development Report 1993* (United Nations, 1993), pp. 163–7.
10. Ibid.
11. Simon Head, 'The New Ruthless Economy', *New York Review of Books* (29 February 1996).
12. Ibid.
13. 'Killer Capitalists', *Newsweek* (26 February 1996).
14. UNDP, *Human Development Report 1993* (New York, Oxford University

Press, 1993), p. 35; Barnet and Cavanagh, p. 294.

15. UNDP, 1993, p. 37.

16. *Global Outlook 2000* (New York: United Nations Publications, 1990).

17. Ibid.

18. 'The Future Tech is Now', *Time* (July 1995).

19. 'Low Commodity Prices', *Financial Times*, 3 February 1999.

4 The Search for El Dorado

1. *Declaration on the Right to Development* (Resolution 41/128 of the United Nations General Assembly, 1976).

2. UNDP, *Human Development Report 1997* (New York: Oxford University Press, 1997), p. 106.

3. Susan George and Fabrizio Sabelli, *Faith and Credit: The World Bank's Secular Empire* (Penguin Books, 1994). See report by Stuart Holland in the *Economist* cited by George and Sabelli, pp. 89–90.

4. Ibid.

5. Thomas Homer-Dixon, Jeffrey Boutwell and George Rethgens, 'Environment, Scarcity and Violent Conflict', *Scientific American* (February 1994).

6. United Nations, *Global Outlook 2000* (UN Publication, 1990).

7. UNCTAD, *Report on Trade and Development, 1996* (New York: United Nations, 1996), pp. 146–9.

8. Michael Pettits, 'The Liquidity Trap', *Foreign Affairs* (November–December, 1996).

9. UNDP 1997, op. cit., pp. 146, 158, 161, 178, 194, 196, 198.

10. Ibid.

11. Ibid.

12. WTO, *Statistics Annual Report 1999* (New York: United Nations, 1998).

13. Eric Hobsbawm, 'Towards the New Millennium', *Sunday Times Review* (16 October 1994).

14. Michel Hirsh, 'Capital Wars', *Newsweek* (3 October 1994).

15. Felix Rohatyn, 'World Capital: The Need and the Risks', *New York Review of Books* (July 1994).

5 Worldwide Depredation

1. Eibl Eibersfeldt, quoted in Lyall Watson, *Dark Nature* (Hodder and Stoughton, 1995), pp. 124, 172.

2. Claude Julien, 'Le Siècle des Extrêmes. Leçons d'Histoire' (The Century of Extremes: Lessons of History), *Le Monde Diplomatique, Manière de Voir* (26 May 1995).

3. Watson, pp. 229–32, 264.

4. Julien.

5. United Nations, *World Investment Report, 1995* (New York: Oxford University Press, 1996), p. 8.

6. Oswaldo de Rivero, 'United Nations: A Peacemaker Without a Sword', *Geneva Post* (May 1995).

7. John Keegan, *A History of Warfare* (New York: Alfred A. Knopf, 1993), pp. 228, 231, 233, 334, 343.

8. Richard J. Barnet and John Cavanagh, *Global Dreams* (New York: Simon and Schuster, 1994), p. 16.

9. Robert Hase, Emily Hill and Paul Kennedy, 'Pivotal States', *Foreign Affairs* (January–February 1996), p. 33.

6 Survival

1. *The Carrying Capacity Briefing Book* (2,600 pp. In 2 volumes) (Washington, DC: Carrying Capacity Network, 1997). Joel E. Cohen, *How Many People Can the Earth Support?* (W.W. Norton, 1996). Donella Meadows, *Beyond the Limits* (Chelsea Green, 1992). Jack Goldstone, *Revolution and Rebellion in the Early Modern World* (Berkeley: University of California Press, 1991).

2. Lester Brown, 'Food Output has Stopped Keeping Up', *International Herald Tribune* (July 26, 1993).

3. Ibid.

4. 'Empty Nets', *Newsweek* (24 April 1994).

5. UNDP, *Human Development Report, 1997* (New York: Oxford University Press, 1997).

6. UN *Special Report on the Social Situation in the World* (United Nations, 1993).

7. Ibid.

8. Ibid.

9. Vaclav Smil, *China's Environmental Crisis* (Armonk, NY: M.E. Sharpe, 1993).

10. *Global Outlook 2000* (United Nations, 1989).

11. UNDP 1997, op.cit.

12. World Resources Institute, *World Resources 1997* (New York: Oxford University Press, 1997).

13. United Nations, *World Urbanisation Prospects: The 1994 Revision* (UN, 1995). 'The Exploding Cities of the Developing World', *Foreign Affairs* (January–February 1996).

14. United Nations, *World Urbanisation Prospects*.

15. UNDP 1997, op. cit.

16. United Nations Population Fund, *State of the World Population 1992* (New York: United Nations).

17. Ibid.

18. Jean Christopher Rufin, *L'Empire et les Nouveaux Barbares* (The Empire and the New Barbarians) (Paris: Collection Pruriel, 1991).

SELECT BIBLIOGRAPHY

An Agenda for Peace, Report of the Secretary General of the United Nations A/47/272-S/24111, 17 June 1992.

Aron, Raymond, *Les Désillusiones du Progrès* (The Disillusions of Progress), Paris: Calmann Levy, 1969.

Barnet, Richard J. and John Cavanagh, *Global Dreams*, New York: Simon and Schuster, 1994.

Berger, Peter L., *Pyramids of Sacrifice*, Pelican Books, 1977.

Brown, Lester, 'Food Output has Stopped Keeping Up', *International Herald Tribune*, 26 July 1996.

Bull, Hedley, *The Anarchical Society*, Macmillan, 1980.

Carrying Capacity Briefing Book (2 volumes), Washington, DC: Carrying Capacity Network, 1997.

Chevallier, J. J., *Grandes Textos Políticos* (Great Political Texts), Madrid, 1962.

Cleveland, Harlan, *Birth of a New World*, San Francisco: Jossey-Bass Publishers, 1993.

Cohen, Joel E., *How Many People Can the Earth Support?*, New York: W. W. Norton and Company, 1996.

Connelly, Matthew and Paul Kennedy, 'Must it Be the Rest Against the West?', *Atlantic Monthly*, December 1994.

Coote, Belinda, *The Trade Trap*, Oxfam, 1993.

Deutsch, Karl W., *The Analysis of International Relations*, Prentice-Hall, 1978.

Drucker, Peter F., *Post-Capitalist Society*, Harper Business, 1994.

Drucker, Peter F., 'The Changed World Economy', *Foreign Affairs*, Spring, 1996

Elliot, Michael, 'The Failed Nations', *Newsweek*, 3 January 1994.

Emerson, Tony, 'Empty Nets', *Newsweek*, April 1994.

Food and Agriculture Organisation, *The Sixth World Food Survey*, FAO, 1996.

Foreign Affairs 75th Anniversary Issue, *The World Ahead*, September–October 1997.

Frye, William, *A United Nations Peace Force*, Carnegie Endowment for International Peace, 1957.

Fukuyama, Francis, *The End of History and the Last Man*, Penguin Books, 1992.

Géopolitique du Chaos (The Geopolitics of Chaos), *Le Monde Diplomatique, Manière de Voir*, February 1997.

George, Susan and Fabrizio Sabelli, *Faith and Credit: The World Bank's Secular Empire*, Penguin Books, 1994.

Gever, John, *Beyond Oil: The Threat to Food and Fuel in the Coming Decade*, Colorado University Press, 1991.

Gilland, Stephen and David Law, *The Global Political Economy*, New York: Harvester Wheatsheaf, 1988.

Global Outlook 2000, United Nations Publications, 1989.

Godoy, Horacio, 'La Crisis del Estado Nacional Contemporáneo' (The Crisis of the Contemporary National State), *América Latina 2001*, Bogota, 1976.

Goldstone, Jack, *Revolution and Rebellion in the Early Modern World*, Berkeley: University of California Press, 1991.

Gray, John, *Post-Liberalism*, London: Routledge, 1996.

Hase, Robert, Emily Hill and Paul Kennedy, 'Pivotal States', *Foreign Affairs*, January/February 1996.

Head, Simon, 'The New Ruthless Economy', *New York Review of Books*, 29 February 1996.

Helman, Gerald and Steve Ratner, 'Saving Failed States', *Foreign Policy 89*, Winter 92–93.

Hirsh, Michel, 'Capital Wars', *Newsweek*, 3 October 1994.

Hobsbawm, Eric, 'Towards the New Millennium', *Sunday Times* Review, October 1994.

Homer-Dixon, Thomas, Jeffrey Boutwell and George Rethgens, 'Environment, Scarcity and Violent Conflict', *Scientific American*, February 1994.

Horsman, Matthew and Andrew Marshall, *After the Nation State*, Harper Collins, 1994.

Huntington, Samuel P., 'The Clash of Civilisations', *Foreign Affairs*, Summer, 1993.

Huntington, Samuel P., 'The West and the World,' *Foreign Affairs*, November– December 1996.

Jackson, Robert H., *Quasi States, International Relations and the Third World*, Cambridge University Press, 1990.

Julien, Claude, 'Le Siècle des Extrêmes. Leçons d'Histoire' (The Century of Extremes: Lessons of History), *Le Monde Diplomatique, Manière de Voir*, 26 May 1995.

Kaplan, Robert, 'The Coming Anarchy', *Atlantic Monthly*, January 1994.

Kaplan, Robert D., 'The Ends of the Earth', New York: Random House, 1996.

Kapstein, Ethan B., 'Workers and the World Economy', *Foreign Affairs*, July–August 1996.

Keegan, John, *A History of Warfare*, New York: Alfred A. Knopf, 1993.

Kenichi, Ohmae, 'Rise of the Region State', *Foreign Affairs*, Spring 1993.

Kennedy, Paul, *The Rise and Fall of the Great Powers*, Random House, 1987.

Kennedy, Paul, *Preparing for the Twenty-first Century*, Glasgow: Harper Collins, 1991.

Keohane, R. O. and Joseph S. Nye, Jr., *Transnational Relations and World Politics*, Harvard University Press, 1966.

Krugman, Paul, 'Emerging Market Blues', *Foreign Affairs*, July–August 1995.

'Le Bouleversement du Monde' (World Upheaval), *Le Monde Diplomatique, Manière de Voir*, 25 February 1995.

Le Désordre des Nations (The Disorder of Nations), *Le Monde Diplomatique, Manière de Voir*, 21 February 1994.

Lellouche, Pierre, *Le Nouveau Monde* (The New World), Paris: Grasset, 1992.

Lemonick, Michael D., 'The Future Tech is Now', *Time*, July 1995.

Linden, Eugene, 'Exploding Cities of the Developing World', *Foreign Affairs*, January–February 1996.

MacMillan, John and Andrew Linklater, *Boundaries in Question*, London, New York: Pinter, 1995.

Minton-Beddoes, Zanny, 'Why the IMF Needs Reform', *Foreign Affairs*, May–June 1995.

Ormerod, Paul, *The Death of Economics*, London: Faber and Faber, 1995.

Parsons, Anthony, *From Cold War to Hot Peace*, London: Michael Joseph, 1995.

Passel, Peter, 'The Poor Get Poorer in the US', *International Herald Tribune*, 30–31 March 1996.

Pettits, Michael, 'The Liquidity Trap', *Foreign Affairs*, November–December 1996.

Porter, Gareth and Janet Welsh, *Global Environmental Politics*, Westview Press, 1996.

Porter, Michael E., *The Comparative Advantage of Nations*, Macmillan Press, 1990.

Powell, Colin, 'Challenges Ahead', *Foreign Affairs*, Winter 1992–93.

'Reconstructing Nations and States', *Daedalus*, Summer 1993.

Reich, Robert B. *The Work of Nations*, Simon and Schuster, 1993.

Rejai, Mostafa and Cynthia H. Enloe, *A Perspective in World Politics*, London: Croom Helm, 1982.

Rohatyn, Felix, 'World Capital: The Need and the Risks', *New York Review of Books*, 14 July 1994.

Rufin, Jean Christopher, *L'Empire et les Nouveaux Barbares* (The Empire and the New Barbarians), Paris: Collection Pruriel, 1991.

'Scenarios de la Mondialisation' (Scenarios of Globalisation), *Le Monde Diplomatique, Manière de Voir 32*, November 1996.

Scheetz, Thomas, *Peru and the IMF*, University of Pittsburgh Press, 1986.

Schwarzenberger, Georg, *State Bankruptcy and International Law*, Deventer, Antwerp, London, Frankfurt, New York: Liber Amicorum Maarten Bros., Wybo P. Herre, Kluwer Law and Taxation Publishers, 1989.

Smith, Adam, *The Wealth of Nations*, Penguin Classics, 1986.

Smith, Michael, Richard Little and Michael Shackleton, *Perspectives on World Politics*, London: Croom Helm, 1981.

Strange, Susan, *Casino Capitalism*, Britain: Basil Blackwell, 1986.

'The Future of the State', *Economist,* issue entitled *A Survey of the World Economy*, September 1977.

The Military Balance 1997–98, International Institute for Strategic Studies, Oxford University Press, 1997.

United Nations, *Reports on the Social Situation in the World*.

United Nations, *UNCTAD World Investment Report*, 1995.

United Nations, *UNDP Human Development Reports*, 1994–97.

United Nations, *UNCTAD Reports on Trade and Development*, 1994–97.

United Nations, *1995 World Urbanisation Prospects: The 1994 Revision*, UN, 1995.

United Nations, *1997 State of the World Population*, UN Population Fund, 1997.

United Nations, *1996 Energy Statistics Yearbook*, New York.

United Nations, *1996 World Population Prospects 1950–2050*.

Watson, Lyall, *Dark Nature*, Hodder and Stoughton, 1995.

World Bank, *World Development Reports 1994–97*.

World Bank Outlook, *Major Primary Commodities*.

World Resources Institute, *World Resources 1997*, New York: Oxford University Press.

World Trade Organisation, *Statistics Annual Report 1999* (WTO Publications, 1998).

Worldwatch Institute, *State of the World 1998* (London: Earthscan, 1998).

INDEX

The Global Issues *Series*

Already Available

Robert Ali Brac de la Perrière and Franck Seuret, *Brave New Seeds: The Threat of GM Crops to Farmers*

Oswaldo de Rivero, *The Myth of Development: The Non-viable Economies of the 21st Century*

Nicholas Guyatt, *Another American Century? The United States and the World after 2000*

Martin Khor, *Rethinking Globalization: Critical Issues and Policy Choices*

John Madeley, *Hungry for Trade: How the Poor Pay for Free Trade*

Riccardo Petrella, *The Water Manifesto: Arguments for a World Water Contract*

In Preparation

Peggy Antrobus and Gigi Francisco, *The Women's Movement Worldwide: Issues and Strategies for the New Century*

Amit Bhaduri and Deepak Nayyar, *Free Market Economics: The Intelligent Person's Guide to Liberalization*

Julian Burger, *Indigenous Peoples: The Struggle of the World's Indigenous Nations and Communities*

Graham Dunkley, *Trading Development: Trade, Globalization and Alternative Development Possibilities*

Joyeeta Gupta, *Our Simmering Planet: What to do about Global Warming?*

John Howe, *A Ticket to Ride: Breaking the Transport Gridlock*

Susan Hawley, *Corruption: Privatization, Multinational Corporations and the Export of Bribery*

Calestous Juma, *The New Genetic Divide: Biotechnology in the Age of Globalization*

John Madeley, *The New Agriculture: Towards Food for All*

Jeremy Seabrook, *The Future of Culture: Can Human Diversity Survive in a Globalized World?*

Harry Shutt, *A New Globalism: Alternatives to the Breakdown of World Order*

David Sogge, *Give and Take: Foreign Aid in the New Century*

Keith Suter, *Curbing Corporate Power: How Can We Control Transnational Corporations?*

Oscar Ugarteche, *A Level Playing Field: Changing the Rules of the Global Economy*

Nedd Willard, *The Drugs War: Is This the Solution?*

For full details of this list and Zed's other subject and general catalogues, please write to: The Marketing Department, Zed Books, 7 Cynthia Street, London N1 9JF, UK or email Sales@zedbooks.demon.co.uk

Visit our website at: http://www.zedbooks.demon.co.uk

PARTICIPATING ORGANISATIONS

• **Both ENDS:** A service and advocacy organisation which collaborates with environment and indigenous organizations, both in the South and in the North, with the aim of helping to create and sustain a vigilant and effective environmental movement.
Damrak 28-30, 1012 LJ Amsterdam, The Netherlands
Phone: +31 20 623 08 23   Fax: +31 20 620 80 49
Email: info@bothends.org
Website: www.bothends.org

• **Catholic Institute for International Relations (CIIR):** CIIR aims to contribute to the eradication of poverty through a programme that combines advocacy at national and international level with community-based development.
Unit 3 Canonbury Yard, 190a New North Road, London N1 7BJ, UK
Phone +44 (0) 20 7354 0883   Fax   +44 (0) 20 7359 0017
Email: ciir@ciir.org
Website: www.ciir.org

• **Corner House:** The Corner House is a UK-based research and solidarity group working on social and environmental justice issues in North and South.
PO Box 3137, Station Road, Sturminster Newton, Dorset DT10 1YJ, UK
Tel: +44 (0)1258 473795   Fax: +44 (0)1258 473748
Email cornerhouse@gn.apc.org
Website: www.cornerhouse.icaap.org

• **Council on International and Public Affairs (CIPA):** CIPA is a human rights research, education and advocacy group, with a particular focus on economic and social rights in the USA and elsewhere around the world. Emphasis in recent years has been given to resistance to corporate domination.
777 United Nations Plaza, Suite 3C, New York, NY 10017, USA.
Tel: 212 972 9877   Fax: 212 972 9878
Email: cipany@igc.org
Website: www.cipa-apex.org

• **Dag Hammarskjöld Foundation:** The Dag Hammarskjöld Foundation, established 1962, organises seminars and workshops on social, economic and

cultural issues facing developing countries with a particular focus on alternative and innovative solutions. Results are published in its journal, *Development
Dialogue*.

Övre Slottsgatan 2, 753 10 Uppsala, Sweden.
Tel: 46 18 102772   Fax: 46 18 122072
e-mail: secretariat@dhf.uu.se
web site: www.dhf.uu.se

• **Development GAP:** The Development Group for Alternative Policies is a
non-profit development resource organisation working with popular organizations in the South and their Northern partners in support of a development
that is truly sustainable and that advances social justice.

927 15th Street, NW - 4th Floor
Washington, DC 20005 - USA
Tel: + 1-202-898-1566   Fax: +1 202-898-1612
E-mail: dgap@igc.org
Website: www.developmentgap.org

• **Focus on the Global South:** Focus is dedicated to regional and global
policy analysis and advocacy work. It works to strengthen the capacity of
organisations of the poor and marginalised people of the South and to better
analyse and understand the impacts of the globalisation process on their daily
lives.

C/o CUSRI, Chulalongkorn University, Bangkok 10330, Thailand
Tel: +66 2 218 7363   Fax: + 66 2 255 9976
Email: Admin@focusweb.org
Website: www.focusweb.org

• **Inter Pares:** Inter Pares, a Canadian social justice organisation, has been
active since 1975 in building relationships with Third World development
groups and providing support for community-based development programmes. Inter Pares is also involved in education and advocacy in Canada,
promoting understanding about the causes, effects and solutions to poverty.

58 rue Arthur Street, Ottawa, Ontario, K1R 7B9 Canada
Tel: + 1 (613) 563-4801   Fax: + 1 (613) 594-4704

• **Third World Network:** TWN is an international network of groups and
individuals involved in efforts to bring about a greater articulation of the
needs and rights of peoples in the Third World; a fair distribution of the
world's resources; and forms of development which are ecologically sustainable

and fulfil human needs. Its international secretariat is based in Penang, Malaysia.

228 Macalister Road, 10400 Penang, Malaysia
Tel: +60-4-2266159   Fax: +60-4-2264505
Email: twnet@po.jaring.my
Website: www.twnside.org.sg

• **Third World Network–Africa:** TWN–Africa is engaged in research and advocacy on economic, environmental and gender issues. In relation to its current particular interest in globalization and Africa, its work focuses on trade and investment, the extractive sectors and gender and economic reform.

2 Ollenu Street, East Legon, P O Box AN19452, Accra-North, Ghana.
Tel: +233 21 511189/503669/500419 Fax: +233 21 51188
email: twnafrica@ghana.com

• **World Development Movement (WDM):** The World Development Movement campaigns to tackle the causes of poverty and injustice. It is a democratic membership movement that works with partners in the South to cancel unpayable debt and break the ties of IMF conditionality, for fairer trade and investment rules, and for strong international rules on multinationals.

25 Beehive Place, London SW9 7QR, UK
Tel: +44 20 7737 6215   Fax: +44 20 7274 8232
E-mail: wdm@wdm.org.uk
Website: www.wdm.org.uk

THIS BOOK IS AVAILABLE IN THE FOLLOWING COUNTRIES:

FIJI
University Book Centre
University of South Pacific
Suva
Tel: 679 313 900
Fax: 679 303 265

GHANA
EPP Book Services
P O Box TF 490
Trade Fair
Accra
Tel: 233 21 773087
Fax: 233 21 779099

MOZAMBIQUE
Sul Sensacoes
PO Box 2242,
Maputo
Tel: 258 1 421974
Fax: 258 1 423414

NEPAL
Everest Media Services
GPO Box 5443, Dillibazar
Putalisadak Chowk
Kathmandu
Tel: 977 1 416026
Fax: 977 1 250176

PAPUA NEW GUINEA
Unisearch PNG Pty Ltd
Box 320, University
National Capital District
Tel: 675 326 0130
Fax: 675 326 0127

RWANDA
Librairie Ikirezi
PO Box 443, Kigali
Tel/Fax: 250 71314

TANZANIA
TEMA Publishing Co Ltd
PO Box 63115
Dar Es Salaam
Tel: 255 51 113608
Fax: 255 51 110472

UGANDA
Aristoc Booklex Ltd
PO Box 5130, Kampala Road
Diamond Trust Building
Kampala
Tel/Fax: 256 41 254867

ZAMBIA
UNZA Press
PO Box 32379
Lusaka
Tel: 260 1 290409
Fax: 260 1 253952

ZIMBABWE
Weaver Press
P O Box A1922
Avondale, Harare
Tel: 263 4 308330
Fax: 263 4 339645